AF896195

Development and Testing of Vehicle Software and its Influence on Sustainable Transport

Antonio Colmenar-Santos · David Borge-Diez ·
Pedro-Miguel Ortega-Cabezas
Editors

Development and Testing of Vehicle Software and its Influence on Sustainable Transport

Springer

Editors
Antonio Colmenar-Santos
Control, Electronics and Electrical Engineering
Universidad Nacional de Educación a Distancia
Madrid, Spain

David Borge-Diez
Electrical, Systems and Automation Engineering
University of Leon
León, Spain

Pedro-Miguel Ortega-Cabezas
UNED
Madrid, Spain

ISBN 978-3-031-47629-7 ISBN 978-3-031-47630-3 (eBook)
https://doi.org/10.1007/978-3-031-47630-3

© The Editor(s) (if applicable) and The Author(s), under exclusive license to Springer Nature Switzerland AG 2024

This work is subject to copyright. All rights are solely and exclusively licensed by the Publisher, whether the whole or part of the material is concerned, specifically the rights of translation, reprinting, reuse of illustrations, recitation, broadcasting, reproduction on microfilms or in any other physical way, and transmission or information storage and retrieval, electronic adaptation, computer software, or by similar or dissimilar methodology now known or hereafter developed.
The use of general descriptive names, registered names, trademarks, service marks, etc. in this publication does not imply, even in the absence of a specific statement, that such names are exempt from the relevant protective laws and regulations and therefore free for general use.
The publisher, the authors, and the editors are safe to assume that the advice and information in this book are believed to be true and accurate at the date of publication. Neither the publisher nor the authors or the editors give a warranty, expressed or implied, with respect to the material contained herein or for any errors or omissions that may have been made. The publisher remains neutral with regard to jurisdictional claims in published maps and institutional affiliations.

This Springer imprint is published by the registered company Springer Nature Switzerland AG
The registered company address is: Gewerbestrasse 11, 6330 Cham, Switzerland

Paper in this product is recyclable.

Preface

"*Development and Testing of Vehicle Software and its Influence on Sustainable Transport*" aims to show the link between software and sustainable transport models and how the software can contribute to increasing the performance of these models and techniques such as Vehicle-to-Building or Vehicle-to-Grid. One of the main concerns is if sustainable transport models are viable economically for all actors involved such as aggregators, the transport system operator, and electric vehicle owners. This book includes a macroeconomic study to show that a sustainable transport model based on electric mobility is viable. Next, as software is an essential topic because it contributes to driving efficiency, smart charging, and eco-routing, a description of the automotive software development and testing processes, as well as novel validation techniques based on expert systems and genetic algorithms, are included in the book. A novel algorithm based on neural networks and the Here® application program interface is described in the book. This algorithm assesses the increasing performance of electric mobility models thanks to eco-driving, eco-routing, and smart charging process. In addition to this, it is shown how vehicle-to-building and vehicle-to-grid performance is increased. Finally, the contribution to the European Green Deal of eco-driving, eco-routing, and smart charging based on the novel algorithm proposed is assessed.

Dordrecht, Netherlands

Anu Peter

Contents

Editors and Contributors

About the Editors

Antonio Colmenar-Santos has a Ph.D. in Industrial Engineering and a Master of Science in Industrial Engineering, with a specialization in Electronics and Automation Engineering awarded both by The School of Industrial Engineering at National Distance Education University (UNED); and a Bachelor of Science in Electrical Engineering, with a specialisation in Electronic Instrumentation, Regulation and Control and Industrial Automation awarded by The School of Industrial Engineering at the University of Valladolid. He has been part of the Spanish section of the *International Solar Energy Society* (ISES) and of the *Association for the Advancement of Computing in Education* (AACE), working in different projects related to renewable energies and multimedia systems applied to teaching. He has been a coordinator of both virtualisation and telematic Services at ETSII-UNED, as well as deputy head teacher (administration) and Head of the Department of Electrical, Electronics and Control Engineering at UNED.

David Borge-Diez has a doctoral degree and a master in industrial technologies research from UNED (Spanish University for Distance Education) and a bachelor's degree in industrial engineering with majors in energetic engineering from the University of Valladolid, Spain. He is a specialist engineer in energy efficiency, energy economics and alternative energy sources and works as a Professor in the Department of Electrical, Control and Automation Engineering in the University of León, Spain. He has been Associated Professor in the University of León and worked for energy companies in both public and private projects, including international R&D programs.

Pedro-Miguel Ortega-Cabezas has a doctoral degree and a Master in Electrical, Electronic and Control Engineering from UNED (Spanish University for Distance Education). Nowadays, he is doing his Ph.D. in Industrial Engineering at UNED. His field of specialization is focused on powertrains and the validation of the engine control unit software. He has participated in several engine design projects launched by such market leaders as PSA Peugeot Citroën and Renault in their design centers located in France. More specifically, his research is focused on how to use artificial intelligence when validating embedded software such as the engine control unit one.

Contributors

Jorge-Juan Blanes-Peiró Department of Electrical and Control Engineering, Universidad de León, Campus de Vegazana, s/n, 24071 León, Spain

David Borge-Diez Department of Electrical and Control Engineering, Universidad de León, León, Spain

Antonio Colmenar-Santos Departamento de Ingeniería Eléctrica, Electrónica y de Control, UNED, Madrid, Spain

Juan-Vicente Míguez-Camiña Departamento de Ingeniería Eléctrica, Electrónica y de Control, UNED, Madrid, Spain

Pedro-Miguel Ortega-Cabezas Departamento de Ingeniería Eléctrica, Electrónica y de Control, UNED, Madrid, Spain

Chapter 1
Macroeconomic Impact, Reduction of Fee Deficit and Profitability of a Sustainable Transport Model Based on Electric Mobility: Case Study—City of León (Spain)

Antonio Colmenar-Santos, David Borge-Diez, Pedro-Miguel Ortega-Cabezas, and Juan-Vicente Míguez-Camiña

Abbreviations

CNE	National Energy Commission
EB	Electric bus
EREN	Regional energy entity
ET	Electric taxi
EU	European electric taxi Union
GDP	Gross domestic product
GPS	Global position system
IPC	Consumer price index

Extract from *Energies* **2020**, *13*(15), 3997; https://doi.org/10.3390/en13153997, https://www.sciencedirect.com/journal/energy.

A. Colmenar-Santos (✉) · P.-M. Ortega-Cabezas · J.-V. Míguez-Camiña
Departamento de Ingeniería Eléctrica, Electrónica y de Control, UNED, Juan del Rosal, 12, Ciudad Universitaria, 28040 Madrid, Spain
e-mail: acolmenar@ieec.uned.es

P.-M. Ortega-Cabezas
e-mail: pedro-miguel.ortega-cabezas@valeo.com

J.-V. Míguez-Camiña
e-mail: jmiguez@ieec.uned.es

D. Borge-Diez
Department of Electrical and Control Engineering, Universidad de León, Campus de Vegazana, S/N, 24071 León, Spain
e-mail: david.borge@unileon.es; dbord@unileon.es

© The Author(s), under exclusive license to Springer Nature Switzerland AG 2024

A. Colmenar-Santos et al. (eds.), *Development and Testing of Vehicle Software and its Influence on Sustainable Transport*, https://doi.org/10.1007/978-3-031-47630-3_1

RE Renewable energy
SGs Smart grids
V2G Vehicle to grid

1.1 Introduction

According to a Eurostat report published in February 2013, EU (European Union) energy dependency in 2011—understood as the relationship between net imports and gross energy consumption—reached 54%. Between 2008 and 2011, consumption fell in 23 of the 27 member states of the EU. The most significant reductions took place in Lithuania (24.5%), Ireland (12.3%), Greece (12.3%), Romania (10.2%), Spain (9.9%) and the United Kingdom (9.9%). The Spanish energy dependency index is 76.4%; 22.6 points above the EU average [1].

The breakdown of final energy consumption by source shows the current energy dependency on non-renewable energies as well as forecasts for 2020, where oil products and derivatives will continue to play a predominant role in the economy (Fig. 1.1) [2].

The transport sector accounts for almost 40% of the total primary energy consumption. This figure can be broken down by transport mode as follows: 80% by road vehicle, 14% by air, 3% by train and 3% by sea. Thus, the sector consumes 65% of the total annual amount of petrol imported by Spain [2].

Considering the trend in terms of absolute variability of energy consumption, the tendency has been dissimilar. It followed the same pattern as GDP (gross domestic

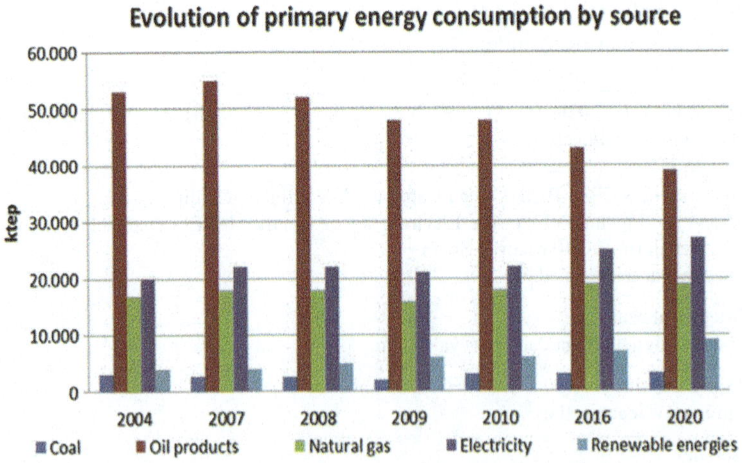

Fig. 1.1 Breakdown of primary energy consumption by source. *Source* IDAE (Institute for Diversification and Energy Saving)

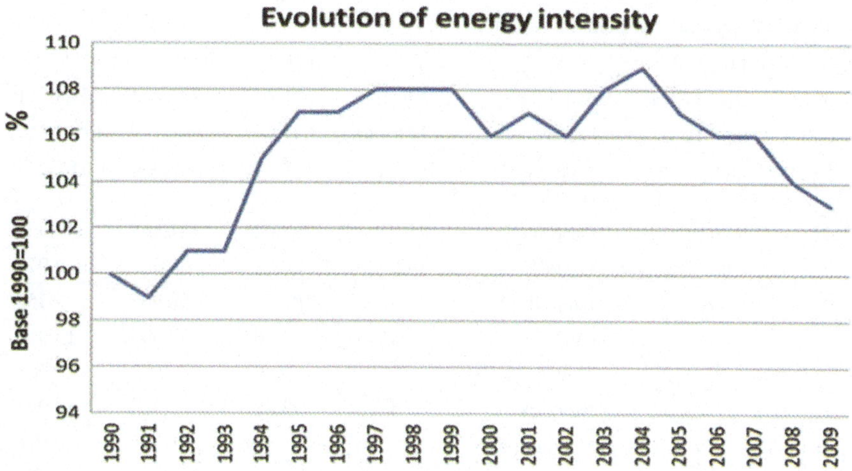

Fig. 1.2 Growth of energy intensity (2004–2010). *Source* Institute for Diversification and Energy Saving

product) from 2005 until 2008, 2009 and 2010, when the energy intensity began to reduce owing to high energy prices and reduction in economic activity. The energy intensity in this area has had a downward tendency since 2004 as a result of lower activity levels in certain sectors and lower mobility of freight and passenger transport in all transport modes (Fig. 1.2) [2].

In Spain, 24% of greenhouse gases (namely CO_2) come from the transport sector. Therefore, significant efforts must be made to achieve the goals established in the various international agreements [2]. According to data published by the EU, the quantified emission limitation or reduction commitment as agreed in accordance with article 4(1) of the Kyoto Protocol for Spain is 115% (percentage of base year or period) [3].

León (42.5984° N, 5.5719° W) is located in the region of Castilla y León (Spain). Its energy consumption pattern does not differ from the rest of Spain. This fact is demonstrated in the 2011 indicator recently published by the EREN (Regional Energy Entity), of the Castilla y León Government [4], as depicted by Table 1.1.

Due to the implementation of energy efficiency measures, combined with the slowing down of economic activity in the main Spanish sectors, a reduction in consumption took place in 2011. This is in line with the rest of the country.

Spain occupies a privileged position in the field of RE (renewable energy) and consequently if their use were encouraged, a significant reduction in Spanish energy dependency would occur. Castilla y León stands out when it comes to wind power and solar energy. Photovoltaic energy has increased from 13 MW installed in 1998 to 5252 MW in 2010. Wind power in this region produces more than 20% of the total generated in Spain, with an installed power of 398 MW. This places Castilla y León in fourth position in relation to the other Spanish regions [5].

Table 1.1 Annual energy demand variation

Conversion factors		Tep	2008	2009	2010	2011
0.086	tep/MWh-PCI	Natural gas	1,646,104	1,621,927	1,782,444	1,763,711
0.086	tep/MWh	Electricity	1,121,980	1,091,957	1,083,169	1,062,349
1.13	tep/t	Liquefied petroleum gas	131,251	113,446	122,359	82,257
1.07	tep/t	Gasoline	417,268	403,026	381,568	349,921
1.035	tep/t	Diesel	3,382,600	3,209,617	3,268,997	3,039,979
0.96	tep/t	Fuel oil	149,156	101,872	68,963	6752
		Coal	4615	2707	30,395	13,905
		Thermal solar	3611	3794	3932	3989
		Biomass	4213	43,681	47,703	49,923
		Geothermal	0	541	726	1016
		Total	6,940,250	6,616,931	6,791,256	6,434,571
Annual variation				4.66%	2.63%	−5.25%

When implementing and developing sustainable urban transport models, different types of factors should be considered; for example, technological, environmental, economic, social, etc. depending on the case-study city. As a result, a generalist model seems unlikely to be proposed, as shown in the conclusions of SUTRA (sustainable urban transportation) [6].

Among the technologies available to implement this model, one can currently find three options: hybrid vehicles, EVs (electric vehicles) and fuel cell vehicles [7–9].

Hybrids are an intermediate case between the other two; they still generate greenhouse gases, but in smaller quantities [10].

As for fuel cell vehicles, they are completely emission-free when operating. As proven by research from the National Renewable Energy Laboratory [11], their performance regarding autonomy and battery life is good. However, hydrogen generation is still a key issue facing problems such as production processes, economies of scale and technological development [12]. Even though this technology is still in development, the possibility of it playing an important role in the future must not be neglected [13, 14].

As for EVs, their main disadvantage is their low autonomy and the insecurity this causes for potential EV (electric vehicle) users [13]. However, this fact should not be considered a major setback for their potential expansion since one of the biggest battery manufacturers in Europe (Axeon) forecasts autonomies higher than 500 km by 2020 [15].

Energy efficiency in the transport sector is a key issue in improving this factor, and many studies are being conducted on this subject. Some of these have aimed to predict energy consumption in this sector on a long-term basis [16, 17]. In addition, some papers have assessed the influence of current European Union policies on private vehicles, especially on the subject of sustainability [18]. Others have focused on the use of electric transport to measure its impact on sustainability [19, 20] as well as a better integration of RE [21, 22]. Further research has centered strongly on the use of quantitative indicators [23].

Implementing these sustainable models brings significant costs as well as income loses for the Government. As a result, their profitability is in doubt.

Previous research [16, 17, 23] has not included specific case studies on sustainable models and their potential uses and economic profitability; instead, they have been limited to environmental aspects and energy dependency. This study has used a traditional model based on electric mobility: SGs (smart grids) [24, 25], EVs (electric vehicles), EBs (electric buses) [26] and ETs (electric taxis). Through a reasoned proposal of adequate remuneration policies, it reconciles the interests of the Government, EV users and generation and distribution companies, for it is only if each of these parties is involved that the success of the proposed plan for a sustainable transport model will be assured. Furthermore, it will allow Spain to not only reduce its energy dependency and environmental impact, but also reduce the current tariff deficit (26,062 MV in March 2013 [27]) without the need to impose high taxes, as has been the case in recent years.

1.2 Liberalization of the Electricity Market. Situation in Spain

All countries where a liberalization of the electricity market has taken place have dealt with imbalances between real costs and regulated fees. The United Kingdom and Eastern European countries have chosen to eliminate the latter, whilst others, such as the USA, have created an extra fee which is added to the regulated one, taking into account generation, transport, distribution and sales costs [28].

However, the Spanish electricity system has not restrained the growth of the fee deficit, with a major increase in costs from 1998 to 2009. Until 2002 this increase was moderate, owing to lower financial costs and greater demand for electricity. Since 2003, the trend has strengthened due to renewable energy production and higher Brent prices [29]. Consequently, the Spanish market suffers from a structural income deficit on regulated activities. This is confirmed by the existing difference between average toll incomes and real system costs according to data published by the CNE (National Energy Commission) (Fig. 1.3) [30].

The Spanish Government has driven several legislative reforms to restrain the fee deficit since 2012. After the approval of RD 1/2012 of 27th January, financial aid to RE facilities for feeding electricity into the grid was suspended. Law 15/2012 of 27th December was passed to impose a fiscal change in the electricity system by introducing three new taxes: one on electricity production value, one on nuclear fuel production and finally one on nuclear radioactive waste. In 2013, RD Law 9/2013 proceeded to suspend financial aid to all RE facilities, even those built before

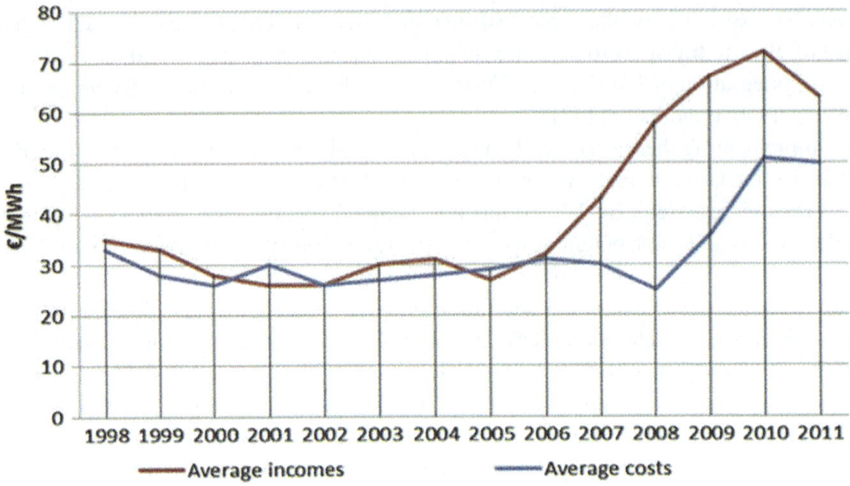

Fig. 1.3 Income and average cost evolution. *Source* CNE

the approval of this RD. Finally, it must be pointed out that tolls have continued to increase in recent years [30].

If the current fee deficit were constrained at its current value, the only way to reduce it would be to increase taxes. As a result, Spanish competitiveness would be strongly affected. Spain currently has the third highest electricity price in the EU [31]. Therefore, this study proposes a sustainable transport model for León by 2020 to contribute to reducing the fee deficit as well as achieving environmental targets, a reduction in petrol consumption and better integration of REs on a large scale.

1.3 Initiatives to Encourage Sustainable Transport

Among the main actions taken to reduce energy dependency in Spain, one can find policies dedicated to encouraging the use of EVs as well as pilot schemes focused on public transport EBs and ETs, which are currently carried out in various Spanish and European cities.

EVs are essential to a sustainable transport model for reducing local CO_2 emissions, due to their performance (85–90%) against traditional vehicles (90%) [32] and their potential for recharging with RE.

The MOVELE Plan (electric mobility plan) is outlined in the Integral Strategy as a means to promote EVs in Spain, between 2010 and 2014. A series of initiatives are being undertaken to stimulate penetration: financial aid for purchasing EVs, the introduction of Charging Manager roles (responsible for selling the necessary electricity to recharge EVs, pushing for the installation of new charge points in public spaces, car parks or shop-ping centers), and the introduction of the peak charge with a view to increasing overnight charging [33].

The final goal is achieving a total of 250,000 EVs in Spain by the end of 2014 [33]. This plan has been brought forward due to cooperation from Movistar [34], who will be in charge of communicating the charge point locations to EV owners through GPS (global position system) and Smartphone applications. As a result, owners will always have information about the nearest charge point. This will help alleviate any concerns about the use of EVs, such as running out of energy mid-journey.

Alongside this plan, the regional strategy strongly advocates EVs in Castilla y León, as led by the regional government. Its main target is achieving 15,000 EVs in Castilla y León [35].

EBs significantly reduce emissions from public transport. There are two factors impeding their widespread use: price and battery life. On the issue of battery life, different pilot schemes were conducted in 2011 and 2012, with significant results. In Madrid, from March 2012, an EB (electric bus) called FOTON was in operation, in order to test performance. The initial data obtained show a saving of 150 V/day/bus in petrol and maintenance [36]. Additionally, from January 2011, the first stage of the PILAVESA project started, in Pamplona (Navarra e Spain). The results obtained after 119 days of operation were: total energy consumption of 142,233 kWh and average daily consumption of 119 kWh. The second stage took place in April 2012; placing solar panels on the roof of the bus with the aim of increasing total battery life over 20–30 km [37]. In June 2012 an EB (model BYD K9), started operating in Barcelona (without passengers). The provisional results are acceptable considering the high traffic intensity in Barcelona. The EB was able to travel 210 km (compared to 250 km indicated by the manufacturer) [38].

Finally, the model considers ETs. In Spain, there is only one type (Nissan Leaf), registered in 2011. The most important results are: 47,600 km traveled, with a daily average of 140 km at 80% capacity, and a standard working day consumption of 13 kWh per 100 km [39].

1.4 Sustainable Transport Model Profitability Study. Case-Study: City of León (Castilla y León-Spain)

When evaluating the profitability of a sustainable transport model, it is necessary to analyze a number of subjects such as the specifics of León, the model itself, the most realistic EV market penetration against Government forecasts, and future energy consumption to assess the share of V2G technology. Finally, a study is carried out on factors such as inflation, costs and energy consumption to establish the impact of this model from an economic, environmental and energy point of view.

1.4.1 Specifics of León

León is a city of 135,059 inhabitants located in the region of Castilla y León (Spain) where considerable investment is being made in the development of SGs. This involves the potential for use of this kind of network together with electric mobility and V2G technology. León is ideal for carrying out this case-study for several reasons.

When studying the current situation of León, it is essential to analyze the use of public and private transport. The first mode to consider is buses. The average age of the fleet is roughly 4 years and incorporates technologies such as Euro IV and biodiesel engines. When it comes to coverage, not all zones of the city are equally covered as not all areas are well connected by bus routes. Taxis are another significant element, with a total of 200 licenses (1.04 taxis/1000 inhabitants). In relation to private transport, it has been estimated using data published in León's urban mobility plan that 101,166 journeys a day are made, with an occupation rate per vehicle of 1.26 people, an average of 3.18 km/journey and an average speed of 11.62 km/h [40].

Tourist attractions and areas for eating out are located in the city center (Fig. 1.4). Because of this, several local by-laws prohibit traffic in certain zones. The rest of the zones are subject to restrictions for two reasons: improvements in inhabitants' quality of life and encouraging tourism. Downtown León has large parking areas which could be used to install charge points. Above all, the objective is to decrease the energy necessary for the operation of the city's transport.

Fig. 1.4 Map: León city center. *Source* Google Maps and León Town Hall

1.4.2 Sustainable Transport Model

In 2007, León had a total of 14 bus lines. Following approval in 2009 of the 'Integral Plan for sustainable mobility in León' this was reduced to 10, in order to reduce petrol. The number of buses necessary to operate public transport will remain constant in the years to come [40]. This study has allowed for 35 EBs to guarantee service.

Considering these figures, the average consumption was 41.5 l/100 km, with a total of 1,817,312 km/year [40].

The transport model described here consists of replacing the current conventional buses with an equivalent number of electric ones (BYD brand equipped with a 324 kWh battery) [41], at a cost of 400,000 V/unit [42]. Despite this, leasing has been chosen instead of buying, following Barcelona's example, with a total cost of 11,580 V/month/bus, maintenance included [43], with the option to buy for V125,750 each in the third year. Annual maintenance is estimated at 0.021 V/km/bus [44]. As a result, this concept will incur a cost of approximately 41,000 V/year, updated annually using the IPC (consumer price index). This amount will be paid from the third year.

The model is completed by 200 ETs operating in the city by 2020. This number has been chosen by the authors since the number of licenses has changed slightly over the last few years and the mobility plan lacks a clear objective on this subject. A slow daily recharge, overnight charge, and quick charge between 1 and 3 pm have been assumed in order to guarantee the battery life required for a full working day. The cost of ETs (Nissan Leaf), which goes up to 24,000 V/unit, has not been considered in the profitability study.[1] This is justified by the cost of its diesel counterpart (approximately 23,300 V/unit in Spain [45]), an assumed 10-year lifespan of 300,000 km, an average consumption of 8.5 l/100 km and a diesel price of 1.4 V/l. All these factors lead to a profit of 35,700 V, without considering income from taxi services, and so the slight original overprice of ETs will clearly be paid off. In addition, savings will be even higher as diesel prices will increase between 2020 and 2030.

Finally, the transport model will include 900 charge points at a cost of 2000 V each [46].

The use of this model implies benefits for all parties involved in the electricity sector as well as León's inhabitants:

a. Electricity companies and aggregators

   Additional income and new business opportunities as a result of electricity sales from V2G, as well as electricity generation for the charging of EBs, ETs and EVs.

b. León inhabitants

   León's inhabitants benefit from the sustainable transport model as the reduction of greenhouse gases such as NO_x, CO_2, and SO_x, will contribute to environmental and well-being improvements.

c. EV owners and the Government

[1] The profitability study can be downloaded from sciencedirect.com.

EV users receive a reduction in maintenance and journey costs compared to traditional vehicle users, in addition to the payment received for joining the grid. However, the Government loses out through decreased income from hydrocarbon and VAT (value added tax) taxes, which reduces the model's economic viability. Despite these considerations, after implementing this model together with retributive and tax policies, the Government is able to collect sufficient revenue to balance the aforementioned losses. As a result, energy dependency, as well as the fee deficit, is reduced, assuring the economic sustainability and competitiveness of the Spanish electricity sector. This study presents a solution which assures profitability for all players involved in the electricity sector.

1.4.3 Situation of Smart Grids in Castilla y León

Currently, different initiatives are being conducted in Castilla y León to implement SGs based on smart counters and distribution networks. IBERDROLA is renewing its distribution network through development of the STAR project. The action plan consists of modernizing all the facilities by 2018, with an investment of 180 MV. This project starts in Salamanca, with an initial budget of 9 MV, replacing all traditional counters with smart ones and modifying a total of 315 power station buildings to adapt them to SGs. In the second stage, from 2013 to 2017, the project will continue in Burgos and León. Upon completion, IBERDROLA will reach a total of 1.6 million supply points with smart counters and 15,200 power station buildings will be adapted to SG technology [47].

1.4.4 EV Penetration Forecast for León

When estimating EV market penetration by 2020, an analysis of population and vehicle fleet growth in León is required. Regarding the former, the figures provided by the INE (National Statistics Institute) imply a decrease from 2006 to 2011. The average population during 2006–2011 was 135,236 inhabitants [48]. Despite the slight rise in 2012, the estimates from the Castilla y León Government for the period 2013–2020 show a population decrease, reaching 4.9% in 2020, as compared to 2012 [49]. Due to the lack of data for León, the authors have assumed the regional tendency. Thus, the average decrease will reach 0.7%/year (Fig. 1.5).

The population estimate for 2020–2030 is based on the demographic projections made for Spain by the INE [50], showing a progressive decline in inhabitants in the coming decades (Table 1.2). Considering these figures, a decrease of 2.6% in the decade 2020–2030 has been assumed (Table 1.3).

In relation to the growth of León's vehicle fleet, the most recent data published in 2010 by the DGT (traffic agency) show an average growth of 2.46% [51]. Due to the lack of data from this date, the authors have assumed possible growth linked to

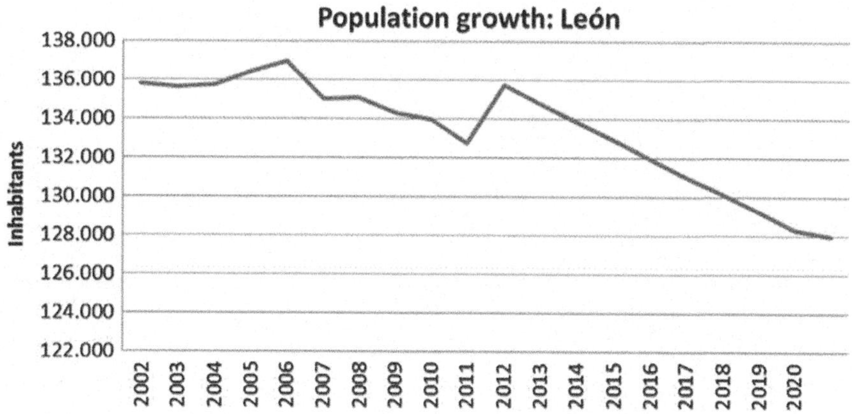

Fig. 1.5 Population growth: León. Estimate made using data from the Government of León

Table 1.2 Spanish population growth

Year	Population	Population variation	
		Absolute	Relative (%)
2012	46,196,278	–	–
2022	45,058,581	−1,137,696	−2.46
2032	43,819,837	−1,238,745	−2.75
2042	42,771,150	−1,048,687	−2.39
2052	41,558,096	−1,213,053	−2.84

Source INE

demographic variability, owing to direct correlation between the use of vehicles and the population (Fig. 1.6).

Various departments and managers directly linked to the automotive sector disagree on the contribution of EVs to Spain's total vehicle fleet by 2020. On the one hand, the Government of the Basque Country estimates it to be 10% [52], whereas the CEO (chief executive officer) of Peugeot Spain sets the figure at 20% (EVs plus hybrids) [53]. In 2030, 2,500,000 EVs are predicted to be driven in Spain, based on a population of roughly 43 million inhabitants [54]. As a result, the ratio that links the number of vehicles to inhabitants is 0.582. Thus, León's contribution is close to 7270 EVs, based on a population of 125,028 inhabitants [48].

The Spanish Government estimates a total of 600,000 EVs by 2020. Based on a figure of 131,556 inhabitants in León and approximately 45,000,000 in Spain, León should contribute 1754 EVs to this objective.

Only one EV was sold up until 2011 [55]. However, the Castilla y León Government predicts this number to reach 15,000 by 2015. Based on a population for Castilla y León of 2,518,683, and 132,913 for the city of León [48], the contribution of the latter should be 790 EVs, a huge increase on the current situation. Hence the authors

Table 1.3 Forecasted population growth

Year	2020	2021	2022	2023	2024	2025	2026	2027	2028	2029	2023
Inhabitants	131,556	131,214	130,872	130,532	130,193	129,854	129,517	129,180	128,844	128,509	128,175

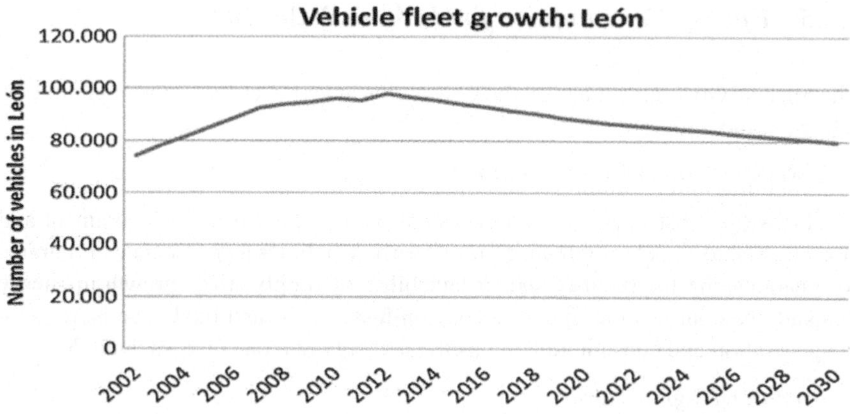

Fig. 1.6 Vehicle fleet growth: León. Estimate made using INE and DGT data

have assumed, for the profitability calculation, that vehicle fleet penetration (Fig. 1.7) will be as follows: from current low values, penetration will increase progressively until it reaches 7270. This figure corresponds to the Government's objective and will be achieved through technological maturity and new models on the market. These estimates are far from the current situation. This disparity of figures could lead this project to be loss-making in the early years and it is therefore necessary to set a worst-case scenario (Fig. 1.7). Despite this, profitability is assured as proven in the Results section.

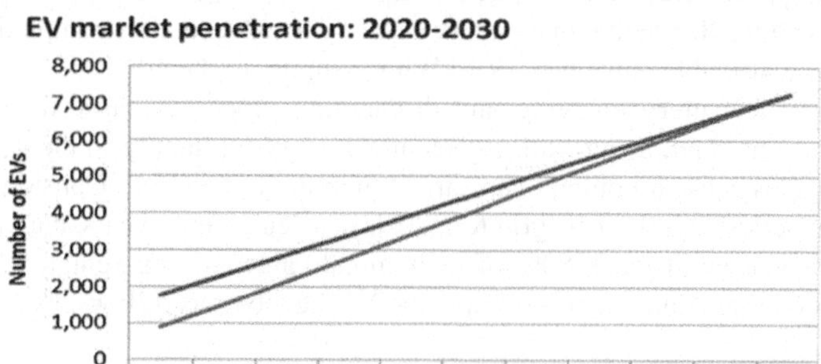

	2020	2021	2022	2023	2024	2025	2026	2027	2028	2029	2030
Number of Evs in worst case scenario	877	1,518	2,159	2,800	3,441	4,082	4,763	5,404	6,045	6,686	7,270
Number of Evs in Government scenario	1,754	2,307	2,861	3,415	3,969	4,523	5,077	5,621	6,175	6,729	7,270

Fig. 1.7 EV market penetration: 2020–2030

1.4.5 Energy Consumption for León in 2020–2030

The Spanish Government has set three scenarios linked to energy consumption: low, reference and high [56]:

a. Reference (or central) scenario

This is the most likely scenario as established by the Spanish Government and considers factors such as retaining current energy policies and associated programs for guaranteeing the security and sustainability of supply, a 2% growth in energy demand, the continuation of dependency on fossil fuels, and finally, an increase in the generation and consumption of electrical energy at a rate of 2.4% until 2030.

b. Low and high scenario

The forecasts for both scenarios are influenced by several factors such as international market growth, demographics, economic development and the environment. Electricity demand will increase at a rate of 3.4%/year (high scenario) and 1.5%/year (low scenario). The Spanish Government is still working with these scenarios. However, taking into account the economic situation and current energy consumption, the low scenario would be the most likely one.

An EV, equipped with a 24 kWh battery (Nissan Leaf [57]), drove a 32 km circuit in downtown León at rush hour, registering accelerations and speeds using a data logger. The aim of this exercise was to estimate the total amount of electricity for supply to the grid (Fig. 1.8). The distance reflects the average daily km driven according to data published by the MIET (Ministry for industry, energy and tourism). Stops and accelerations due to traffic intensity increase the consumption of EVs in Wh/km. Hence the final battery capacity is reduced after completion of the journey. The available value was 19.1 kWh for supply to the grid.

The battery charging and discharging processes cause a decrease in battery capacity and, as a result, in potential energy for supply to the grid, as every 1,500 cycles cause a nominal capacity reduction of 20% [58]. Considering that an EV is expected to link to the grid for 220 days a year, after 6 years, the maximum available capacity will reach 80% of its nominal value. As the vehicle fleet gets older, the forecasted daily contribution to the V2G evolves as follows, Eq. (1.1):

$$p_{\mathrm{V2G}} = n_{80\%}.f.p + (d - n_{80\%}).p \tag{1.1}$$

where $n_{80\%}$ represents the number of EVs with maximum battery capacity after recharging of 80% of its nominal value, n is the number of EVs in the considered year, f is the decreasing factor of the battery capacity due to discharging and charging cycles (20% considered) and p is the estimated energy to be fed into the grid (considered to be 0.0191 MWh).

The results are shown in Table 1.4.

In relation to ETs and EBs, no contribution will be made at any time. In fact, fast charges should be made to assure battery capacity lasts a working day.

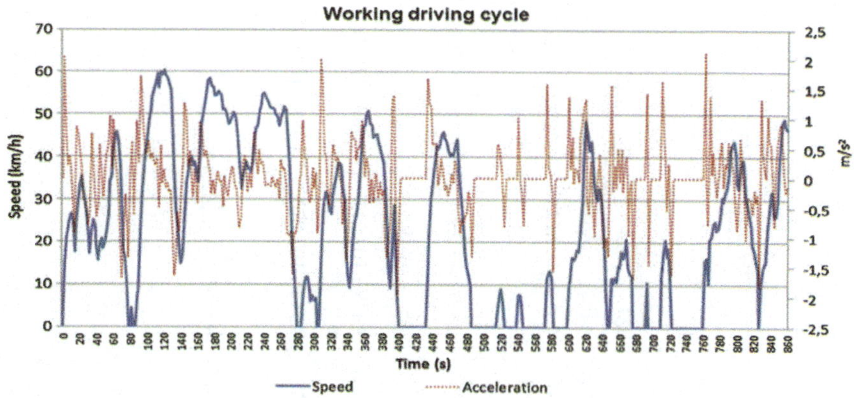

Fig. 1.8 Speeds and accelerations registered during working day (Empirical data obtained using a data logger and an electric vehicle carrying out a 32 km journey in León. This number of km will not change for two reasons: Leon's population trend and the fact that expansion of the city has already been completed. As a result, the size of León will remain the same)

The average daily consumption (MWh) estimated in 2020 for León would be 159 MWh. The Spanish TSO (transport system operator) does not provide data at a local level. Its minimum unit is regional. As a result, this figure was obtained by linking León's projected inhabitants in 2020 (131,556) [48, 49] with those of Castilla y León (2,469,826) [48, 49], to weight the data provided by the TSO.

1.4.6 Income from the EV Charging Process. Losses for Hydrocarbon Tax

The same circuit shown in Fig. 1.8 was conducted at rush hour, measuring consumption with a debimeter, in order to estimate losses due to hydrocarbon tax. The obtained result was 2.048 l. In relation to taxis, the previous result (2.048 l) was extrapolated. The result was 8.704 l, which takes into consideration 136 km/day from the data published by León Town Hall. León's conventional diesel buses consume 2065 l after driving 4978 km/day in 2020, once the reorganization of bus lines indicated in the Leon Mobility Plan [40] has been taken into account.

Estimating these consumptions in euros requires a prediction of the price of oil in 2020. From a euro/dollar parity of 1.2 $/V and the worst-case scenario set by the IDAE (Institute for diversification and energy saving), the barrel price will reach 105.57 V in the reference scenario established by the US DOE (US Department of Energy) [59]. Because its price in May 2013 was 80.85 V/barrel, an increment of 5% a year is assumed from 2013 to 2020. Currently the cost of diesel is 1.348 V/l, and thus, in 2020, 1.82 V/l.

Table 1.4 Worst case scenario and the Government's position on the contribution of EVs to V2G

Year	2020	2021	2022	2023	2024	2025	2026	2027	2028	2029	2030
Number of EVs in worst-case scenario	877	1518	2159	2800	3441	4082	4763	5404	6045	6686	7270
Number of EVs in Government scenario	1754	2307	2861	3415	3969	4523	5077	5621	6175	6729	7270
Number of EVs in worst-case scenario (with reduced battery)	0	0	0	0	0	0	877	1518	2159	2800	3441
Number of EVs in Government scenario (with reduced battery)	0	0	0	0	0	0	1754	2307	2861	3415	3969
% battery reduction	0	0	0	0	0	0	20	20	20	20	20
Energy injected in the grid (MWh) in worst-case scenario	16,751	28,994	41,237	53,480	65,723	77,966	87,623	97,418	107,212	117,007	125,712
Energy injected in the grid (MWh) in Government scenario	33,501	44,064	54,645	65,227	75,808	86,389	90,270	98,548	107,013	115,479	123,695

Note Government data have been analyzed and set out in Sect. 1.4.4
A residual presence of EVs is assumed in 2020. Thus, the possible impact on the model due to charging and discharging processes is neglected

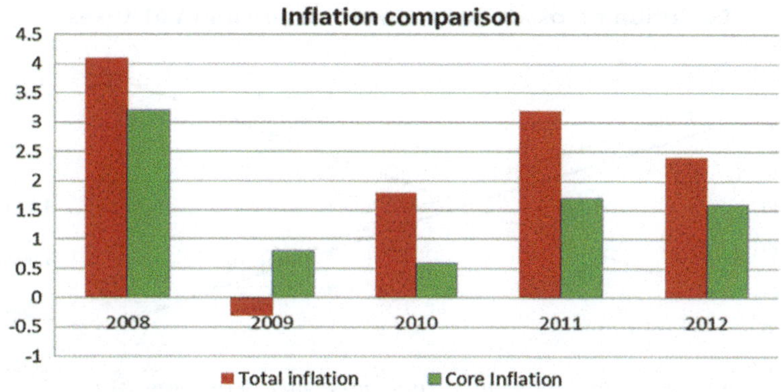

Fig. 1.9 Comparison between core inflation and inflation. *Source* INE

The Government will earn income from EV charging depending on the price of kWh in 2020–2030. Determining this date requires assessment of future inflation and the current forecast of the kWh price in 2020. IDAE assumptions forecast 0.0732 V/ kWh [2]. Regarding inflation, the IMF (International Monetary Fund's) forecasts have been used [60]. Because this organization makes forecasts until 2017, the authors, considering the data published by the IMF until 2017, have established a possible growth between 2020 and 2030 as depicted by the profitability study. However, the Spanish Government has linked electricity with core inflation. This index connects the evolution of prices, removing the effect of energy products and raw commodities. In spite of the fact that there are no forecasts for its trend, its value has been lower than inflation so far [61], except in 2009 when the oil price changed exceptionally (Fig. 1.9). A decrease of 0.8% according to inflation is assumed for this profitability study. Regarding the EV recharging process, this study proposes the EV owner buys the energy at pool price.

Losses due to hydrocarbon tax (42.8%) and its VAT (21%) are growing in line with current policies as EVs increase their penetration in the market. As a result, fiscal modifications are necessary (Fig. 1.10).

1.4.7 Pollution

The diesel vehicle used on the circuit (Fig. 1.8) emits 100 g CO_2/km. Therefore, private cars generate 32 g/km/day and taxis 136 g/km/day.

According to figures published in León's mobility plan, buses will cover 4978 km/ day in 2020 once the reorganization of bus lines has been taken into account [40]. The CO_2 emitted is 0.089 kg CO_2/passenger/km [62]. An average occupancy of 13 people and a 30-strong fleet are assumed.

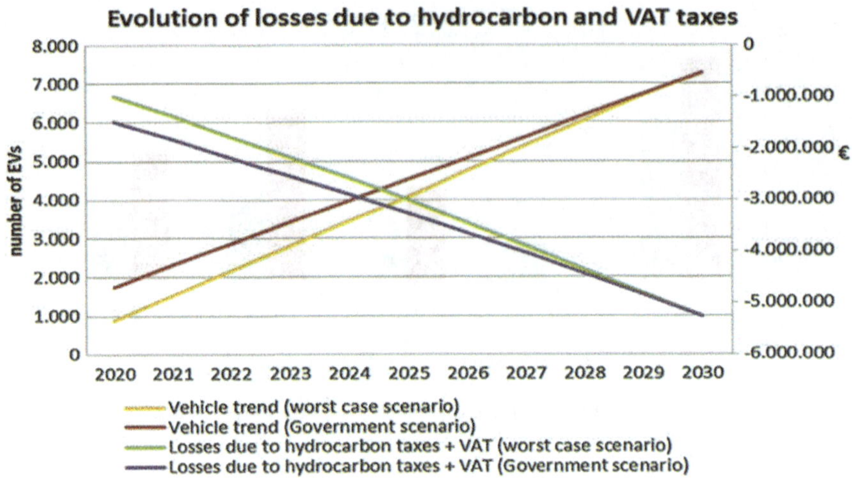

Fig. 1.10 Losses due to hydrocarbon and VAT taxes considering EV penetration

The forecasted price in 2020 is 14 V/t of CO_2 [63], without any expected increase, unless annual CO_2 emissions are reduced. In this case the price could reach 32 V/t of CO_2. In this research, a price of 17 V/t is assumed.

Table 1.6 depicts the results.

1.4.8 Tax and Retributory Policies

Modifications to tax and retributory policies are required to recover the losses of hydrocarbon taxes estimated in Sect. 1.4.6. This research proposes:

a. Electricity tax

When implementing a sustainable transport model, all citizens benefit from environmental improvement, whether or not they own EVs. Therefore, a 0.25% increase on electricity tax is proposed, i.e. from the current value of 4.68–4.93%. This charge is paid by all consumers (even companies). As a result, the increment must be moderate.

b. V2G transaction tax

All incomes from V2G technology must be shared among the different players in the electricity system. When EV owners feed energy into the grid using their battery, they will get paid 67% of the price pool. Aggregators will receive 33% [64]. The incomes are shown in Eq. (1.2) [64]:

$$p_{V2G} = \frac{p_{pool}}{50}.h.p_{consumer} \tag{1.2}$$

Table 1.6 Total emissions and associated taxes

Year	2020	2021	2022	2023	2024	2025	2026	2027	2028	2029	2030
CO_2 emissions per year (220 days considered) in worst-case scenario (t)	5862.70	6007.11	6151.51	6295.92	6440.32	6584.73	6738.14	6882.55	7026.95	7171.36	7302.92
Incomes from sales (17 V/t) in worst-case scenario	99,665.98	102,120.85	104,575.73	107,030.60	109,485.48	111,940.36	114,548.42	117,003.30	119,458.18	121,913.05	124,149.63
Emissions of CO_2 a year (220 days considered) in Government scenario (t)	6060.28	6184.85	6309.66	6434.47	6559.27	6684.08	6808.88	6931.43	7056.24	7181.04	7302.92
Incomes from sales (17 V/t) in Government scenario	103,024.68	105,142.53	107,264.22	109,385.91	111,507.59	113,629.28	115,750.97	117,834.36	119,956.04	122,077.73	124,149.63

where p_{pool} is the price in *V* of MWh in the electricity pool, *h* (taken as 4 h) is the number of hours that users join the grid and $p_{consumer}$ (taken as 67%) is the consumers' benefit percentage. With such shares all parties in the system receive a return on their investment, as proven in the profitability study.

Users' and aggregates' incomes must be subject to deductions by the Government. Aggregators and electricity companies take advantage of these models for several reasons: their business activities grow because of the charging process for V2G technology and their operational costs are reduced as energy is, in certain situations, obtained directly from EVs. This study does not propose to change the current value of corporate incomes and generation taxes, 35% and 7% respectively, since an increase could affect consumers. For users, there are two options. Firstly, the V2G payment can be considered as an additional annual income with the corresponding IRPF (personal income tax) increase, along with the risk of not paying off the battery. Thus, participation in V2G would be viewed negatively. Another option considers incomes arising from V2G being subject to the IRPF but at the same time exempt (i.e. at 0%). This fact will help the Government gather sufficient information about the transactions.

Therefore, this study proposes to create a fee on users' annual incomes for joining the grid. Its value has been set at 5% to assure the Government's return without affecting the battery pay-off by the users.

c. Motor vehicle tax

In Spain, vehicles are levied with an annual tax. EVs, however, are exempt. Therefore, as their market penetration increases, Government losses also increase; and so a balance must be reached [65]. The level of taxation must take three factors into account: consumption-efficiency (kWh/km), power (kW) and size. The Government must establish the weighting of each parameter to guarantee the necessary revenues. An average increase of 10 V/vehicle against current incomes is assumed in this chapter.

d. Registration tax

This tax is paid only once at the time of purchasing an EV. Its level is influenced by the same factors as mentioned previously. A slightly higher increase than that levied on traditional vehicles should be considered. In this research, the increment is set at 50 €/vehicle. Vehicle and registration taxes aim to finance the infrastructure necessary to implement this model, bearing in mind the potential profit for EV owners.

e. Tolls

An increment of 1% is considered to assure profitability.

1.5 Results

This model will be profitable as long as profitability is guaranteed for the Government, EV users, electricity companies and aggregators.

1.5.1 Government

This sustainable transport model based on electric mobility significantly reduces Government income from hydrocarbon tax and VAT (value added tax). Potential levies such as tolls, registration and motor vehicle taxes are supposed to balance the loss of income. However, the profitability associated with this model is achieved by assuring profits for EV owners from V2G and good EV market penetration. Regarding the former, as already exposed, the final owner's profit is close to 7800 V after the battery is paid off. Inarguably these results encourage all EV owners to feed into the grid. Regarding the latter point, there is a key discrepancy between the different Governments' forecasts. After analyzing the growth of León's vehicle fleet and its demographics, a comparison has been made between the situation described by the Government and a more pessimistic scenario, considering the number of EVs sold to date in Castilla y León. In both cases, profitability is assured. In the first scenario, payback is 6 years, TIR (internal rate of return) 20.14%, NPV (net present value) 34,252,068 V and net profits are close to 36,052,068 V between 2020 and 2030. In the second case, payback is 5 years, TIR 14.44%, NPV 31,490,628 V and net profits are close to 33,290,628 V. These figures could be used to decrease the fee deficit of 26,062 MV. Considering that the Spanish population will reach 44,836,892 by 2020 [50], León's fee deficit is 76,468,558 V. In the worst-case scenario, there is a 43.5% reduction compared to 47.1% in the Government scenario.

Profitability is ensured by two taxes:

a. Corporate Income Tax set at 35% of the aggregator and generator companies' profits.
b. 5% V2G Tax, proposed by the authors, on the total income earned by EV owners through the sale of energy to companies.

Increases in road, registration and electricity taxes are very low compared to previous levels and, consequently, their contribution to the final profitability is minor (Figs. 1.11 and 1.12).

Even though these results show significant profitability for the Government, they can fluctuate due to estimation errors regarding tolls, the MWh cost in 2020 and EV market penetration. In fact, tolls are a key issue. As depicted in Fig. 1.13, either they are increased or the model would turn out loss-making. Nonetheless, with increases of 1%, a 20.14% TIR would be obtained in the Government scenario.

MWh price is not a key issue for this model. From a 30% estimation error regarding the value established by the IDAE (73.2 MWh) this project becomes loss-making.

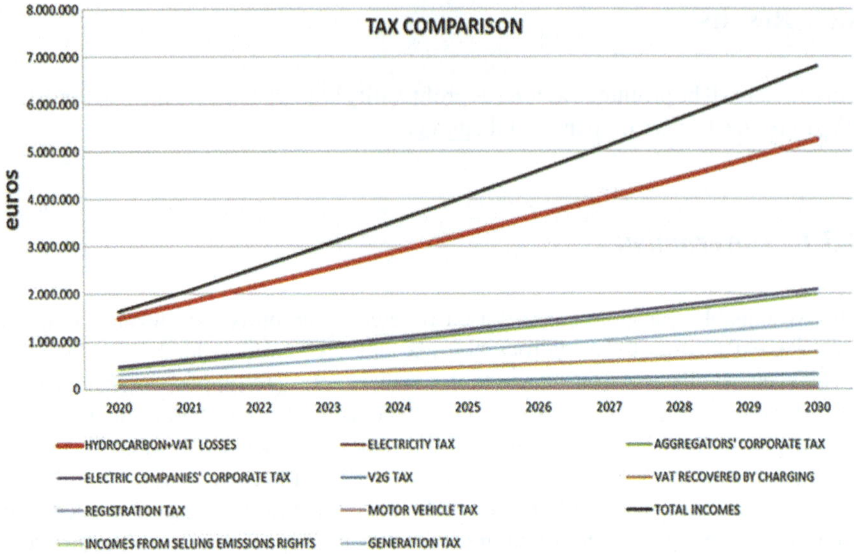

Fig. 1.11 Comparison: taxes against losses in hydrocarbon tax and associated VAT (Government scenario)

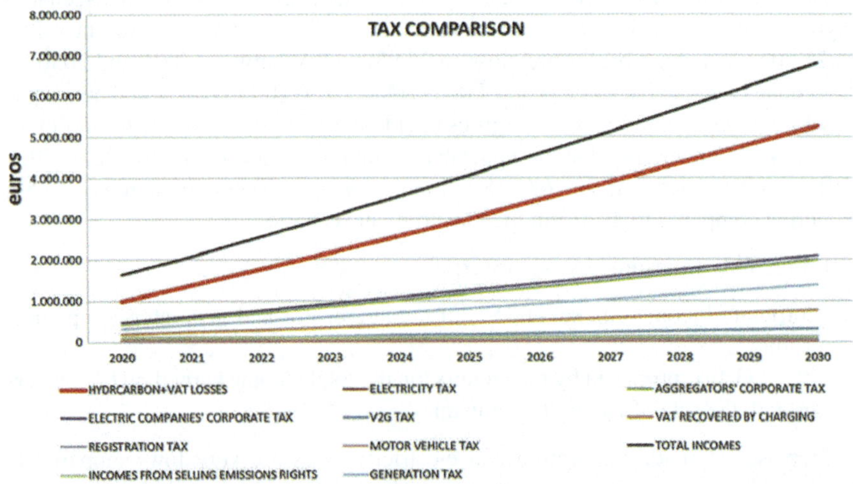

Fig. 1.12 Comparison: taxes against losses in hydrocarbon tax and associated VAT (worst-case scenario)

This inaccuracy implies the electricity price will not increase in 8 years. This fact is unlikely considering the current fee deficit (Fig. 1.14).

This model relies strongly on the level of EV market penetration. If this level is lower than the Government's forecast, the TIR declines (Fig. 1.15). However, even

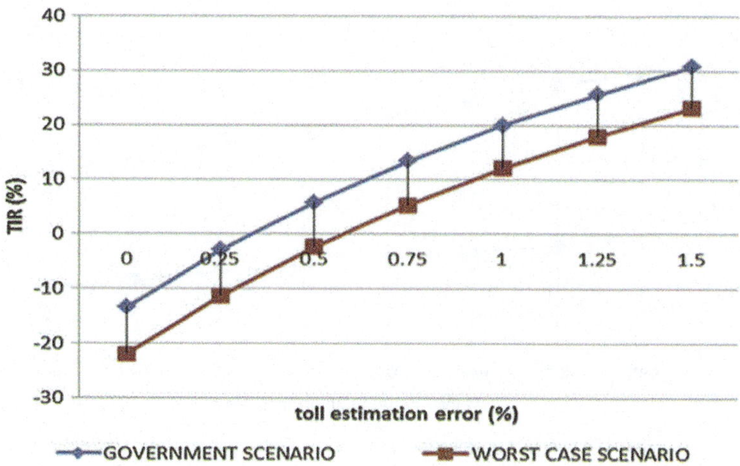

Fig. 1.13 TIR against tolls

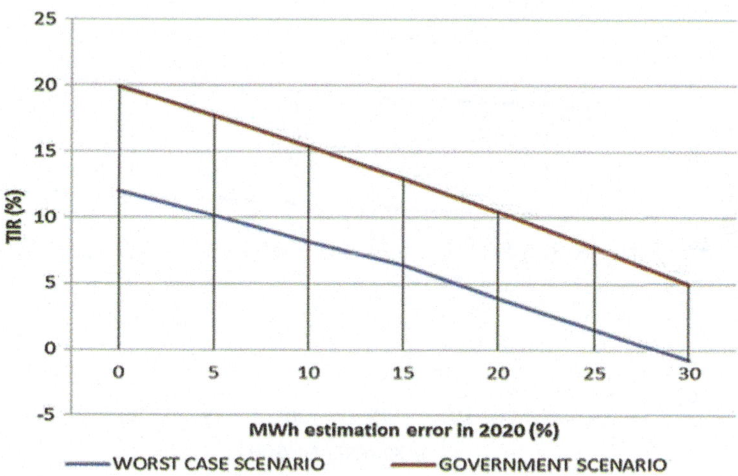

Fig. 1.14 TIR against MWh estimation error

if it reaches 15% and a 5% V2G tax is applied, positive TIRs are still obtained. This estimation error can be put on the same level as V2G user participation.

It is important to point out that the Government's decision to carry out this type of project must also take social benefit into consideration, rather than focus solely on TIR, payback and NPV. In this case both views are positive.

This model does not affect the other taxes mentioned in this study. However, it is important to increase them to implement and maintain new charging points. As examples, Figs. 1.16 and 1.17 show TIR evolution against V2G and motor vehicle taxes.

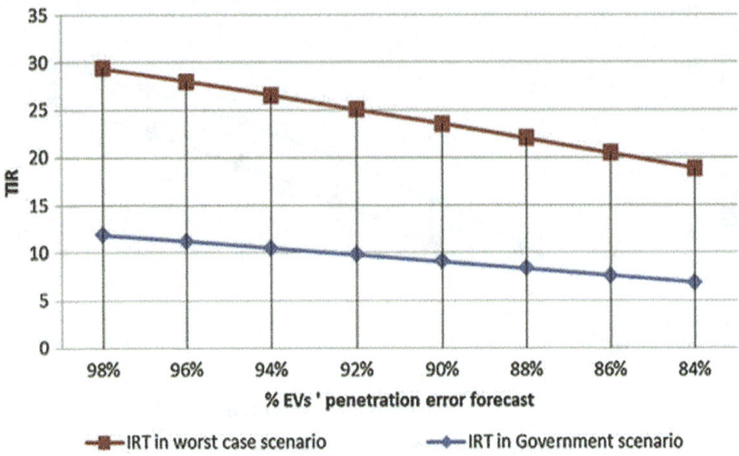

Fig. 1.15 TIR plotted according to the Government's forecast on EV penetration

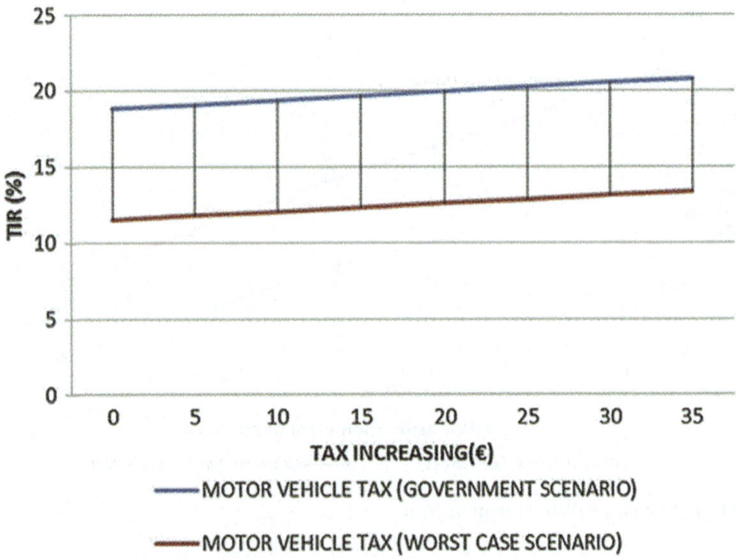

Fig. 1.16 TIR against motor vehicle tax

However, their turnover will be increased due to the charging processes of EVs, ETs and EBs.

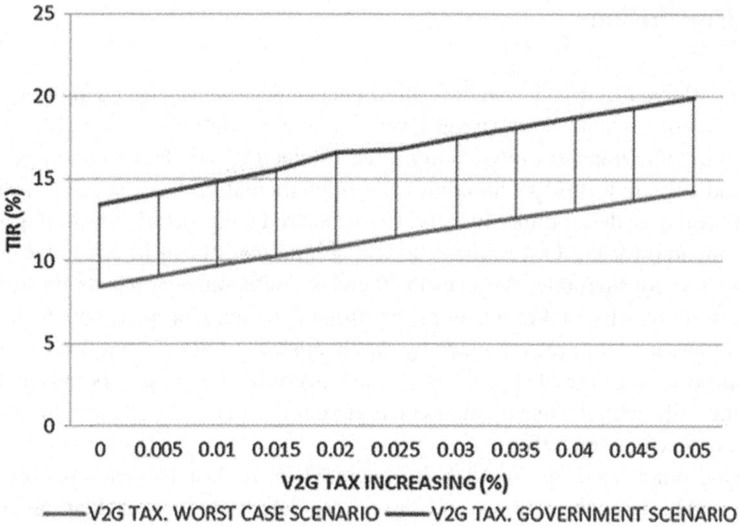

Fig. 1.17 TIR against V2G tax

1.5.2 Aggregators and Electricity Companies

Aggregators will not incur any economic risks as their main function will be selling electricity coming from batteries in the pool and paying users for it. As for electricity companies, taking into account the current overcapacity of the Spanish electricity system, they will not be forced to build new RE facilities (wind power or solar energy).

1.5.3 EV Users

This model will be profitable as long as users get profits. They make money by selling electricity using V2G and through significant diesel savings. On the other hand, their main expenses are motor vehicle tax, registration tax, battery pay-off and charging costs. As proven in the profitability study, a user will get 7800 V once the battery is paid off. This conclusion is supported by various consultants who forecast a cost of 130–200 V/kWh for ion-lithium batteries in 2020–2025. According to the report by McKinsey & Company, EVs will be competitive if battery prices reach 205 V/kWh or less, and as a result EV users will not suffer from higher purchase costs in the future [66].

1.6 Conclusions

Spain is affected by two key issues: high energy dependency and a significant fee deficit, which are both at worrying levels. A single solution can be applied based on a sustainable transport model composed of electric cars, buses and taxis, smart grids and vehicle-to-grid technologies. Its implementation faces several problems, one of which is the commonly held view that reducing petrol consumption will reduce income for the Government. Likewise, the investment to conduct this type of project is considerable. As a result, there is doubt surrounding its profitability. This case study, which takes into consideration the scenario proposed by the Spanish Government and the forecasts made by the Spanish TSO, shows that it is possible to make this project profitable for EV users, the Government, aggregators and electricity companies, through the use of adequate regulatory policies. As a result, losses from petrol taxes will be recovered.

On the other hand, the fee deficit generated in the last 10 years as a result of imbalances between the incomes and costs of regulated activities reached 26,062 M V in May 2013. The Results Section shows that León's fee deficit can be reduced by 43.5–47.1%. In addition, implementation of the model leads to a reduction in petrol demand and, thus, a potential reduction in price. Furthermore, this profit could be used to reactivate financial aid for renewable energy facilities. This fact improves Spanish companies' competitiveness and decreases operational costs for electric vehicles.

As a result, the Spanish economy will enjoy increased protection against petrol price fluctuations. Expected TIR, NAV, payback and profitabilities are sufficient to assure that implementation of these models is considered, as well as providing the means to reduce the fee deficit. The authors are working on preparing future studies focused on rolling out these models to bigger cities such as Madrid and Barcelona.

References

1. Eurostat http://epp.eurostat.ec.europa.eu/cache/ITY_PUBLIC/8-13022013-BP/EN/8-13022013-BP-EN.PDF (last accessed February 2013)
2. Institute for Diversification and Energy Saving. Plan de Acción Nacional de Energías Renovables de España 2011e2020. Ministry of Industry, Energy and Tourism; 2011. http://www.idae.es/index.php/mod.documentos/mem.descarga?file/documentos_11905_PAEE_2011_2020._A2011_A_a1e6383b.pdf (last accessed March 2013)
3. European Union http://ec.europa.eu/clima/policies/g-gas/docs/table_emm_limitation_en.pdf (last accessed October 2013)
4. Government of Castilla y León http://www.energia.jcyl.es/web/jcyl/Energia/es/Plantilla100/1284280587559/_/_/_ (last accessed March 2013)
5. Government of Castilla y León http://www.energia.jcyl.es/web/jcyl/Energia/es/Plantilla100Detalle/1267710822752/_/1284254870263/Comunicacion (last accessed February 2013)
6. SUTRA Project http://www.ess.co.at/SUTRA/ (last accessed October 2013)
7. Gonzalez-Palencia JC, Furubayashi T, Nakata T. Energy use and CO_2 emissions reduction potential in passenger car fleet using zero emission vehicles and lightweight materials. Energy 2012, 48(1), 548–65

8. Zhang Q, Mclellan BC, Tezuka T, Ishihara KN. A methodology for economic and environmental analysis of electric vehicles with different operational conditions. Energy 2013; 61: 118–127.
9. Hedegaard K, Ravn H, Juul N, Meibom P. Effects of electric vehicles on power systems in Northern Europe. Energy 2012, 48(1), 356–68
10. Bradley TH, Frank AA. Design, demonstrations and sustainability impact assessments for plug-in hybrid electric vehicles. Renew Sustain Energ Rev 2009, 13(1), 115–28
11. National Renewable Energy Laboratory http://www.nrel.gov/hydrogen/pdfs/54860.pdf (last accessed October 2013)
12. Lucia U. Overview on fuel cells. Renew Sustain Energ Rev 2014, 30, 164–9
13. Wang D, Zamel N, Jao K, Zhou Y, Yu S, Du Q, et al. Life cycle analysis of internal combustion engine, electric and fuel cell vehicles for China. Energy 2013, 59, 402–12
14. Wasselynck G, Auvity B, Olivier JC, Trichet D, Josset C, Maindru P. Testing of a fuel design and testing of cell powertrain with energy constraints. Energy 2012, 38(1), 414–24
15. Journal: El Pais http://blogs.elpais.com/coche-electrico/2012/04/baterias-con-500-kilometros-de-autonomia-a-partir-de-2020.html (last accessed October 2013)
16. Puksec T, Krajacic G, Lulic Z, Vad Mathiesen B, Duic N. Forecasting long-term energy demand of Croatian transport sector. Energy 2013, 57, 169–76
17. Al-Ghandoor A, Samhouri M, Al-Hinti I, Jaber J, Al-Rawashdeh M. Projection of future transport energy demand of Jordan using adaptative neuro-fuzzy technique. Energy 2012, 38(1), 128–35
18. Aranda-Urs6n A, Valero-Capilla A, Zabalza-Bribián I, Scarpellini S, Llera-Sastresa E. Energy efficiency in transport and mobility from an eco-efficiency viewpoint. Energy 2011, 36(4), 1916–23
19. Hu X, Chang S, Li J, Qin Y. Energy for sustainable road transportation in China: challenges, initiatives and policy implications. Energy 2010, 35(11), 4289–301
20. Smith WJ. Can EV (electric vehicles) address Ireland's CO_2 emissions from transport? Energy 2009, 35(12), 4514–21
21. Jian L. Electric vehicle charging infrastructure assignment and power grid impacts assessment in Beijing. Energy Policy 2012, 51: 554–7
22. Foley A, Tyther B, Calnan P, Gallachoir BO. Impacts of Electric Vehicle charging under electricity market operations. Applied Energy 2013, 101: 93–102
23. Scarpellini S, Valero A, Llera E, Aranda A. Multicriteria analysis for the assessment of energy innovations in the transport sector. Energy 2013, 57: 160–8
24. Malik AS, Bouzguenda M. Effects of smart grid technologies on capacity and energy savings e a case study of Oman. Energy 2013, 54: 365–71
25. Soares B, Borba MC, Szkio A, Schaeffer R. Plug-in hybrid electric vehicles as a way to maximize the integration of variable renewable energy in power systems: the case of wind generation in northeastern Brazil. Energy 2012, 37(1): 469–81
26. Kühne R. Electric busesean energy efficient urban transportation means. Energy 2010, 35(12), 4510–3
27. National Energy Commission. https://www.cnmc.es/sites/default/files/1552698_8.pdf (last accessed February 2013)
28. Energy and Society Web. https://www.energiaysociedad.es/manual-de-la-energia/1-5-el-proceso-de-liberalizacion-de-los-sectoresenergeticos/ (last accessed October 2013)
29. Banco Bilbao Vizcaya Argentaria (BBVA) http://juanst.com/wp-content/uploads/2010/09/Deficit-Tarifario-Espa%C3%B1ol-por-el-BBVA.pdf (last accessed October 2013)
30. National Energy Commission http://www.cne.es/cne/doc/publicaciones/20120309_PI_DEFICIT_ELECTRICO.pdf (last accessed October 2013)
31. Ministry of Industry and Tourism http://www.minetur.gob.es/es-es/gabineteprensa/notasprensa/2013/documents/presentacion_reforma%20el%C3%A9ctrica120713_v5.pdf (last accessed October 2013)
32. Newton T. How cars work. 1st ed., Black Apple Press, United States of America, 2009
33. MOVELE Project. http://movele.es/ (last accessed January 2013)

34. MOVISTAR. https://www.movistar.es/blog/movistar-car/localizador-gps-para-coche-beneficios-y-ventajas-de-movistar-car/ (last accessed April 2013)
35. Government of Castilla y León. Guia del vehículo electrico para Castilla y León; 2011
36. Metropolitan Transport Company of Madrid (EMT). http://www.aemtbus.org/2012/03/la-emt-prueba-el-autobus-modelo-foton/ (last accessed February 2013)
37. Government of Navarra. https://www.navarra.es/home_es/Actualidad/Sala+de+prensa/Noticias/2012/04/26/autobus+electrico+pilavesa.htm (last accessed January 2013)
38. BYD Brand http://www.byd-auto.es/actualidad/22-transporte-electrico/77-el-autobus-electrico-byd-ebus-probado-por-emt-y-tmb.html (last accessed January 2013)
39. Forocoches Web http://forococheselectricos.com/2012/07/nueve-meses-conduciendo-un-taxi.html (last accessed September 2013)
40. León Town Hall http://www.aytoleon.es/es/ayuntamiento/areasmunicipales/urbanismo/Documents/Plan_Director_PMUS_Leon.pdf (last accessed March 2013)
41. BYD Brand. https://bydeurope.com/pdp-bus-coach (last accessed February 2013)
42. Journal: El Periódico http://www.elperiodico.cat/ca/noticias/barcelona/barcelona-timid-primer-pas-cap-lautobus-100-electric-2685671 (last accessed June 2013)
43. Metropolitan Transport Company of Cataluña http://www.tmb.cat/es/sala-de-premsa/-/seccio/noticies/home/noticia-prova-bus-electric-barcelona-20130314-innovacio (last accessed January 2013)
44. U.S. Department of Energy http://www.afdc.energy.gov/pdfs/chatt_cs.pdf (last accessed June 2013)
45. Motor Pasión Web http://www.motorpasionfuturo.com/coches-electricos/electrico-vs-diesel-nissan-leaf-frente-a-volkswagen-golf (last accessed October 2013)
46. Charging Point http://www.chargingbox.es/productos.html (last accessed May 2013)
47. IBERDROLA. https://www.iberdrola.com/about-us/what-we-do/smart-grids (last accessed February 2013)
48. National Statistics Institute. https://www.ine.es/jaxiT3/Tabla.htm?t=2877&L=0 (last accessed May 2013)
49. León's Government region. León en cifras. La población de León; 2011
50. National Statistics Institute. www.ine.es/prensa/np587.pdf (last accessed May 2013)
51. Traffic Agency. https://apps.fomento.gob.es/bdotle/visorBDpop.aspx?i=396 (last accessed March 2013)
52. Journal: El Correo http://www.elcorreo.com/vizcaya/20120627/mas-actualidad/sociedad/cerca-coches-electricos-circularan-201206272006.html (last accessed March 2013)
53. Infocoches http://coches.forumsee.com/a/m/ns/p12-1687-0281264elos-coches-electricos-supondran-las-ventas-europeas-2020.html (last accessed May 2013)
54. General Foundation CSIC http://www.fgcsic.es/lychnos/es_es/articulos/evaluacion_impacto_integracion_coche_electrico (last accessed May 2013)
55. Government of Castilla y León http://www.vehiculoelectrico.jcyl.es/web/jcyl/VehiculoElectrico/es/Plantilla100/1284194828567/_/_/_ (last accessed May 2013)
56. Institute for Diversification and Energy Saving http://www.idae.es/uploads/documentos/documentos_11227_PER_2011-2020_def_93c624ab.pdf (last accessed January 2013)
57. Nissan http://www.nissan.es/ES/es/vehicle/electric-vehicles/leaf.html (last accessed October 2013)
58. Dhameja S. Electric battery systems. 3rd ed., Newness, United States of America, 2002
59. U.S. Energy Information Administration http://www.eia.gov/forecasts/aeo/IF_all.cfm (last accessed October 2013)
60. International Monetary Fund. https://www.imf.org/en/Publications/WEO/Issues/2016/12/31/Legacies-Clouds-Uncertainties (last accessed March 2013)
61. National Statistics Institute http://www.ine.es/daco/daco42/daco421/ipc0012.pdf (last accessed January 2013)
62. Carbon Independent. http://www.carbonindependent.org/sources_bus.htm (last accessed April 2013)

63. OKO Institut: Institute for Applied Ecology. https://www.oeko.de/fileadmin/oekodoc/Emissions-trading-system-for-transport-and-buildings-RefFix.pdf (last accessed May 2013)
64. Beck LJ. V2G-101. 1st ed., BookSurge Publishing, United States of America, 2009
65. Green Car Reports http://www.greencarreports.com/news/1081332_a-reminder-wa-electric-car-owners-pay-100-tax-from-2013 (last accessed April 2013)
66. Motor Pasion web http://www.motorpasionfuturo.com/coches-electricos/el-coste-de-las-baterias-podria-reducirse-hasta-130-kwh-en-2025# (last accessed October 2013)

Chapter 2
Software Validation Techniques in the Automotive Sector

David Borge-Diez, Pedro-Miguel Ortega-Cabezas, Antonio Colmenar-Santos, and Jorge-Juan Blanes-Peiró

Abbreviations

ATCU	Automatic Transmission Control Unit
ADAS	Advanced Driver-Assistance System
dll	Dynamic-link libraries
CAN	Controller Area Network
ECU	Electronic Control Unit
ESP	Electronic Stability Program
EX	Expert system
GA	Genetic Algorithms
HIL	Hardware-in-the-loop
MIL	Model-in-the-loop
SIL	Software-in-the-loop
SM	Software Module

D. Borge-Diez · J.-J. Blanes-Peiró
Department of Electrical and Control Engineering, Universidad de León, Campus de Vegazana, s/n, 24071 León, Spain
e-mail: david.borge@unileon.es; dbord@unileon.es

J.-J. Blanes-Peiró
e-mail: jorge.blanes@unileon.es

P.-M. Ortega-Cabezas · A. Colmenar-Santos (✉)
Departamento de Ingeniería Eléctrica, Electrónica y de Control, UNED, Juan del Rosal, 12, Ciudad Universitaria, 28040 Madrid, Spain
e-mail: acolmenar@ieec.uned.es

P.-M. Ortega-Cabezas
e-mail: pedro-miguel.ortega-cabezas@valeo.com

© The Author(s), under exclusive license to Springer Nature Switzerland AG 2024

A. Colmenar-Santos et al. (eds.), *Development and Testing of Vehicle Software and its Influence on Sustainable Transport*, https://doi.org/10.1007/978-3-031-47630-3_2

2.1 Introduction

2.1.1 Engine ECU Software

Electronic control units (ECUs) have become essential for the correct operation of a vehicle [1, 2]. Software validation plays a key role and has two fundamental goals [3]. Firstly, the software must comply with the functional specifications set by the design team. Secondly, software validation ensures the integration of all software modules (SMs) into the hardware, simultaneously checking that all the elements present in the network interact properly [4, 5]. The process of software validation of an ECU implies significant costs for the companies during a project because of the means necessary to carry out this activity [6, 7]. In addition, the cost of correcting bugs, once the software is marketed, is high and it can tarnish the brand's image [8, 9]. Consequently, a balance between costs, deadlines, and quality must be reached.

Powertrain control is a system in charge of transforming the driver's will into an operating point of the powertrain according to the performance established for the product [10]. The key element of the control system is the engine ECU composed of complex hardware and software. The engine ECU (hardware and software) must be validated to assure that engine is properly controlled, the interaction with the rest of the ECUs is rightly performed and the passengers' safety is insured. Thus, one can deduce that the software validation process is complex and needs improvements with the aim of reducing costs, increasing productivity and reliability in the automotive sector [11, 12].

This chapter is focused on the engine ECU software validation and shows solutions to the main difficulties associated with traditional software validation techniques by using expert systems (EXs) and dynamic-link libraries (dlls) during the hardware-in-the-loop (HIL) simulation. The technique proposed in this research performs better than traditional techniques and allows improving: ease for automating test-cases, bug detection skills, functional coverage, difficulties to detect bugs linked to SMs that do many calculations and the difficulties to validate the software automatically among others. In addition, it shows that the HIL simulation can be automated in an easier way.

2.1.2 Related Works

The code and functional coverage is a real concern when validating a software. Research has been conducted on this topic to enhance this parameter [13–17]. Therefore, test-case generation is a key issue. The black-box technique has been used for a long time in the automotive sector, as discussed by Conrad [18]. Despite its widespread use, it is true that it has some weak points as discussed by Chundur et al. [19]. In their dissertation, they consider that test-cases based on the engineers' experience usually imply gaps and test-redundancies. The model-based testing technique is an option to assess the code and functional coverage rate. The generation and execution of test-cases based on models have been proposed on several occasions. For instance, Skruch and Buchala (DELPHI supplier) proposed a study based on models [20]. The tool Automation Desk (dSpace®) was used. Raffaelli et al presented research focused on functional models by using the commercial software Matelo® [21, 22].

The HIL simulation should be carried out as quickly as possible and with the highest number of test cases executed to ensure the time-frame and quality of the project [23]. Test automation is essential to ensure a high code coverage and to improve reliability [24, 25]. There are many ways for automating HIL simulation in the market [26, 27]. The automation process is mainly based on black-box techniques such as stated by Lemp, Köhl and Plöger: "*As a rule, the tests specified by the ECU departments are first performed as black box tests on the network system (know-how on software structures is not taken)*".

The HIL simulation implies that a specific operating point is reached by the engine ECU. This can be extremely complicated, requiring a lot of manipulations on the HIL model due to SM interactions. There are three possible ways for executing a given test-case in an HIL simulation. Firstly, executing the test-case manually, that is, a technician performs all the necessary actions in the HIL simulation to reach the desired operating point. Secondly, the "tester-on-the-loop" concept can be used. Petrenko, Nguena-Timo and Ramesh, reported the main problems and solutions associated with software validation in the automotive sector [28]. Their main conclusion was focused on the methodology known as "tester-in-the-loop", in which the test engineer leads the system to a desired operation point, considered as a crucial operation point. Once the crucial point is reached, a series of automated actions are executed to reach the goals previously established in the test-case. Finally, test-cases can be fully automated. In this case, a script controls the whole execution process.

Some types of bugs are not detected by using some techniques such as the tester-in-the-loop or black-box, Fig. 2.1, depicts the obtained result for an output for a variable of a SM when executing the software in an HIL simulation (in red) and its expected value (in blue). As one can see, the results are different. This error represents an inaccuracy when it comes to calculating the gas speed in the exhaust pipe. This error impacts the amount of urea injected to treat NO_x. Because this bug does not imply the presence of a functional bug, it is impossible to detect it by using the black-box technique. The detection of this type of bugs involves the checking and detailed analysis of the software code by running additional software.

The solution for validating no matter what type of SM is very far from achieving by employing a direct comparison between the HIL results and the expected outputs indicated in the test-cases. One can encounter some difficulties such as synchronization problems or difficulties to validate the software automatically, among others. Table 2.1 describes the main issues.

The present chapter proposes how to implement the possible solutions depicted in Table 2.1 thanks to the use of dlls for validating any types of SMs when automating a test-case through the HIL simulation, and especially all SMs that cannot be validated

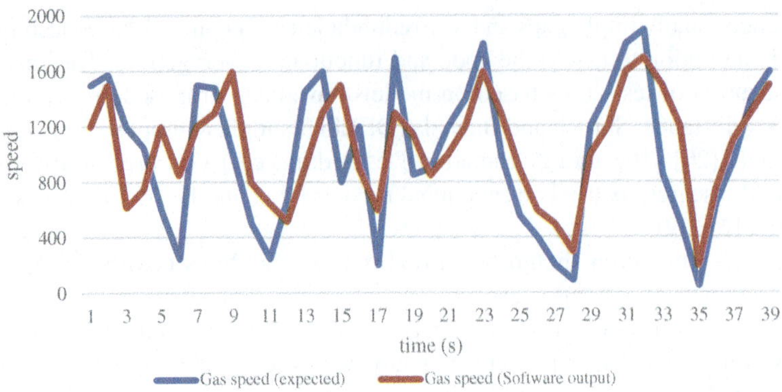

Fig. 2.1 Bug not detected when using traditional techniques

by employing traditional techniques. Thanks to dlls, SMs responsible for doing a great deal of internal calculations, can be validated. During the HIL simulation, it can be checked that all the calculations are properly carried out when the software and hardware are integrated. This feature allows finding bugs which cannot be found by traditional techniques. In addition, in case the desired operating point set in the test-case is not reached in an automated HIL simulation, owing to SM interactions, the dlls can determine the expected output that the software should provide. Thanks to rule-based EX, it is possible to verify whether the functional behavior of the software is correct for the outputs obtained after the HIL simulation. EXs can carry out a real-time performance validation when executing a test-case thanks to dlls.

2.2 Application of Rule-Based Expert Systems and Dynamic-Link Libraries to Enhance Hardware-In-The-Loop Simulation Results[1]

2.2.1 Introduction

New and innovative techniques to validate software are needed to reduce cost and increase software quality.

This research focuses on the validation of engine electronic control unit software by using EXs and dlls with the aim of checking if this technique performs better than traditional ones.

[1] Extract of the following paper published in *Journal of Software* "Application of Rule-Based Expert Systems and Dynamic-Link Libraries to Enhance Hardware-In-The-Loop Simulation Results" JSW 2019 Vol.14(6): 265–292. ISSN: 1796-217X. https://doi.org/10.17706/jsw.14.6.265-292. http://www.jsoftware.us/.

Table 2.1 Potential solutions for the aforementioned issues

Consequences	Reason	Possible solutions
Difficulties to validate the software automatically	When the values set in the test-case for the inputs are not reached due to SM interactions; then the output values set in the test-case may be no longer available. No automatic validation can be performed	Recalculate the output values So that automatic validation process can be carried out. Dlls can perform this task
	The test-engineer cannot establish the expected outputs before performing the test. In some cases, the output values are analog trends which depend on many factors (number of kilometers, number of regenerations of the diesel particulate filter, values of safety module counters, dilution oil rate, properly EEPROM initialized, etc.). Consequently, the expected output can be set after having performed the HIL simulation performing the test. In some cases, the output values are analog trends which depend on many factors (number of kilometers, number of regenerations of the diesel particulate filter, values of safety module counters, dilution oil rate, properly EEPROM initialized, etc.). Consequently, the expected output can be set after having performed the HIL simulation	
Bug performance detection	If input values are different from the ones established in the test-case, then the software performance behavior is unknown	
Synchronization problems	When a test-case is run, the process must compare the current state of the engine ECU and the expected outputs. It is not possible to read all variables involved in the test-case at the same time due to data acquisition software limitations combined with Python scripts Consequently, a desynchronization problem occurs as some variables are read at t1, others at t2 etc.	A data-acquisition can be done while the test-case is run. Then, when the process is ended, the data-acquisition is stopped, and the conformity of the results can be achieved comparing the HIL results with the dll results
	The fact of having different values stored in EEPROM memories keeps the test-engineer from providing accurate screenshot and expected results	The EEPROM can be initialized when building the dll

(continued)

Table 2.1 (continued)

Consequences	Reason	Possible solutions
Functional coverage unknown	A functional code coverage could be established by analyzing the black-box test-cases before the HIL simulation. When reaching different values for the inputs after HIL simulations, then the use-cases tested are different from the ones planned	Implementing a system that can assess whether the software performance is as expected or not Considering the number of performance rules assessed, the functional coverage could be established. A performance EX can perform this task
Difficulties to detect bugs linked to SMs that perform many calculations	The calculations may be performed wrongly but they do not imply that the vehicle behaves in such a way that the client could detect any abnormality (Fig. 2.1)	Implementing a system that can check if the software properly calculates all software outputs. Dlls can perform this task

To do this, a test-case database was built and run by using HIL simulations to validate a series of SMs by using these techniques: the tester-in-the-loop, automation by using a Python script, the model-based testing and EXs combined with dlls with the aim of assessing several factors such as: productivity gain, bug detection skills, functional coverage assessment, ease to automate test-cases among others.

Dlls and EXs improve the HIL success rate by 4.8%, 6% and 20% at least, for simple, fairly-complex, and highly-complex SMs, respectively. Between 9 and 13 more bugs were found when using the EXs and dlls compared with other techniques. Two of the bugs would have required software not initially planned as they were linked to environmental policies. The proposed technique can be applied to any types of a SM, especially in those cases in which traditional validation techniques fail.

2.2.2 *Method*

2.2.2.1 Description

The engine specifications are composed of Simulink® models. Thus, the dll can be easily built considering that Mathworks® has implemented different ways to build a dll from a Simulink® model [29].

The method used in this research are composed of different stages. Firstly, a series of test-cases are designed. Then, all test-cases are run by using the following techniques: manual execution by a technician, automation by employing Python scripts (with and without dlls), the tester-in-the-loop technique and fully automated process by using a performance EX combined with dlls. The EX compiles all rules (software requirements) related to the SM under validation. To conduct the test-cases, an HIL simulation is used. The HIL model belongs to the company subjected to this case-study and has been validated by its experts. The hypothesis to be proved by following this method is that all issues shown in Table 2.1 can be solved thanks to this technique proposed by the authors. Several indicators are analyzed such as: evaluation of the success rate of the HIL simulation, main causes of failure and success for each of the methodologies when running test-cases, the functional coverage obtained, the productivity gain which may take place. The advantages and limitations of using dlls will be discussed. EXs will assess the software performance.

The dll can be implemented by following the steps indicated in many Mathworks® documentation available in their site. The only thing that the user really needs is the Simulink® model to be converted into a dll. In this study, this is not a problem as the specifications needed to code the engine ECU software, are composed of Simulink® models. The main difficulty is how to call the dll. To do this, as described in Matlab® documentation, different programming languages such as C or an m-file can be employed. In this research, C language has been chosen. It is important to describe how the HIL simulation is performed when using dlls to validate the software. Figure 2.2 depicts the process when using an automation script. This description is valid for all techniques but the manual execution one (no automation process). A test-case is executed through a Python script coded by a test engineer. At this moment, the software Inca® [30], or any other software that can read the memory positions of the ECU, performs the data acquisition of all the software variables selected by the test engineer. The result of this process is to generate a data-acquisition file. During the HIL simulation the script is in charge of performing all the necessary manipulations on the driver-ECU interface of the HIL model automatically. If after a certain pre-established time, the values for the input set in the test case are not reached, the data acquisition process and the test-case execution are stopped by the Python script. Then, a data acquisition file containing all the software variables chosen by the test-engineer in the HIL simulation is obtained. A C-file is in charge of decoding the data acquisition file and sending, one by one, all the samples of the HIL simulation to the dll as exposed later. Every time a sample is sent by the C-file, the dll returns the theoretical value that the software should have delivered. Then, the Python scripts

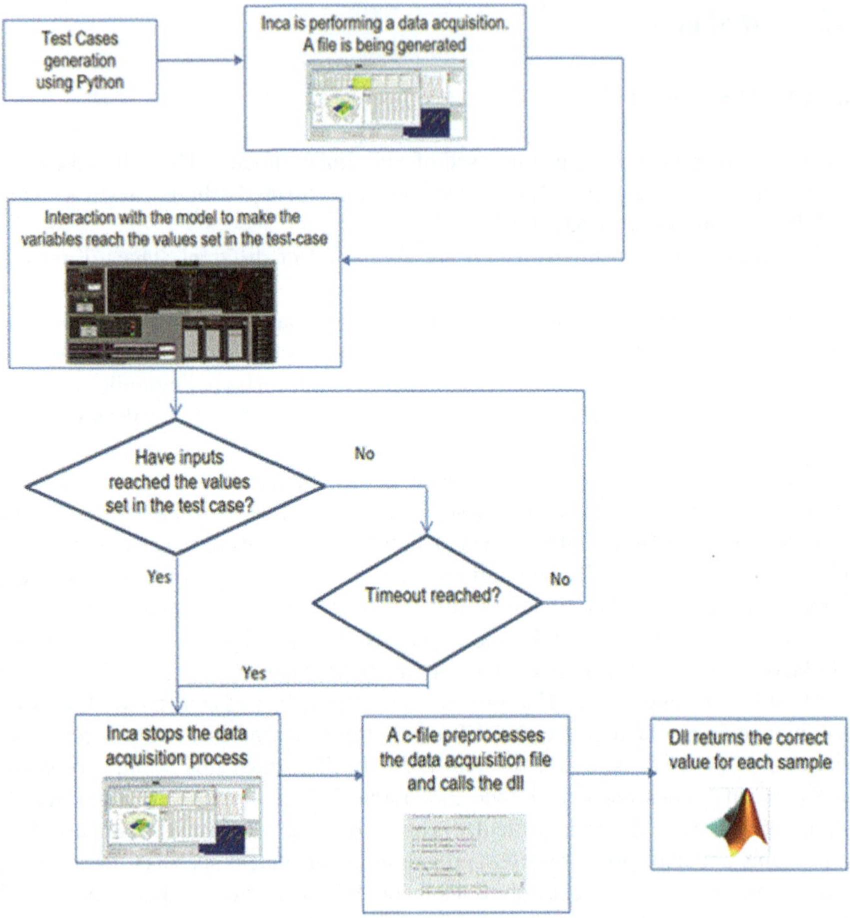

Fig. 2.2 Use of dlls in an HIL simulation when performing a test-case

checks whether the software outputs are equal to dll outputs every time the dll returns a value. Two key topics must be reminded. Firstly, the outputs of the SM are also available in the asci-ii file. Secondly, the engine ECU software is an image of the Simulink® models of the SM under validation.

2.2.2.2 Functions Used in the HIL Simulation

The methodology proposed in this study has been tested in three types of functions or SMs chosen according to the number of calculations to be done as well as their complexity, number of inputs and/ outputs of the SM and the accuracy required for the output results (Table 2.2). They have been considered as representative for this case-study by the authors and the company subjected to this research.

Table 2.2 Types of SM presented in the ECU software

Type of SM	Characteristics	Validation requirements	SM
Simple	a. A reduced number of input and output variables present in the SM and small number of calculations to be done. Furthermore, they are not complex b. High accuracy needed for calculations in some cases and easy to identify the main functional characteristics of the SM	SMs require a few manipulations to make the engine ECU reach the desired operating point For instance, the SM in charge of detecting whether the accelerator pedal is blocked. The engine ECU must check a few parameters	Such as: Temperature estimators Brake pedal monitoring
Fairly complex	a. High number of input and output variables present in the module but moderate number of calculations to be performed b. Moderate accuracy needed for calculations. However, difficult to identify the main functional characteristics of the SM	SMs require more manipulations to make the engine ECU reach the desired operating point For instance, SMs related with treatment of exhaust gases	Such as: Treatment of exhaust gases systems
Highly complex	a. High number of input and output variables and number of calculations b. Calculation not necessarily complex but high number of functional calculations but Moderate/low calculation accuracy	SMs need weeks to reach the desired operating point For instance, the SM in charge of assessing the diesel dilution rate in the engine oil	Such as: The SM in charge of controlling the oil rate diluted into diesel

It is important to establish this classification because the validation requirements as well as the characteristics of the SM clearly influence the time required to carry out the validation process, as well as the additional difficulties that may arise. 5 SMs of each type were selected, based on different criteria such as test engineers' experience, the most problematic SMs in other projects, SMs that require systematic validations to ensure the vehicle safety, SMs that require frequent regression validations as well as those SMs that have never been implemented in previous projects and, in short, they are a novelty (see Table 2.2).

Table 2.3 shows the number of tests considered in this research according to the type of SM.

Table 2.3 Number of tests used in this research

Type of SM	Number of test
Simple	250
Fairly complex	1,250
Highly complex	100

Table 2.4 Methods to generate test-cases

Technique	Method
Cause-effect technique	A1
Model-based testing	A2
One EX combined with dlls and Two EXs combined with dlls	A3

A1: A database in which the staff trace different bugs found throughout a project. In addition, several test-cases come from the software requirements
A2: Pseudorandom values generated by Matelo® to cover a functional model
A3: Pseudorandom values generated by Python scripts

Table 2.4 indicates the methods followed to generate test-cases for each technique.

It is important to analyze what A2 and A3 mean. In A2, Matelo® can generate all necessary test-cases with the aim of covering the functional model. In A3, Python scripts also generate test-cases trying to cover the functional model. In addition, they generate pseudorandom values trying to reach functional states not implemented in the model. A functional state not implemented in the model involves a use-case not considered by the design team. In other words, a design error. The fact of using fuzzy variables, as exposed later, allows increasing the combination of the inputs of the SM under validation. It must also be taken into account that the scripts in charge of generating pseudorandom values have to avoid impossible combinations such as a vehicle speed at 90 km/h and the first shift engaged.

Table 2.5 shows examples of test-cases which could be used to check some functionalities of the software by using different techniques. Fuzzy variables are used when using EXs combined with dlls by increasing the number of combinations of the inputs provided by the SM under validation.

2.2.2.3 Equipment

The following equipment was used in this research.

- An engine ECU software and hardware.
- The HIL bench used to conduct this research belongs to the manufacturer dSpace®, model dSpace® Simulator Full-size [31]. It is a versatile HIL simulator capable of emulating the dynamic vehicle behavior.
- When it comes to building the model that serves as the driver's interface, ControlDesk® version 5.1 from dSpace® manufacturer is employed [32]. By using this software, it is possible to carry out all necessary data exchange between the HIL bench and the engine ECU. This model was designed by the company subjected to this case-study and it is validated by the Electronic Validation Powertrain and Hybrids service before using it.
- Throughout this research, it is necessary to make measurements of different software variables stored in the engine ECU memory. To do this, it is imperative to

Table 2.5 Examples of test-cases

Feature to be checked	Actions to be done	Expected results	Technique
Body control unit. Cyclic redundancy check invalid	Set a CRC invalid value of the frame BCM_A1	Check the inhibition of adaptive cruise control	Cause-effect Model-based testing
Diesel particulate filter regeneration	1. Var1_veh_started = TRUE Start the vehicle 2. Var2_temperature_exhaust_gas = 600°C Do a driving cycle and var3_vehicle_speed = 80 km/h Press the brake pedal to reach 40 km/h Then Var4_particulate_filter = 40 g Do not overpass 2000 rpm	When the RG is performing the variable var1_out is activated	Model-based testing
Diesel particulate filter regeneration	Var1_veh_started = TRUE Start the vehicle and var2_temperature_exhaust_gas = High Do a driving cycle. Var3_vehicle_speed = High Press the accelerator pedal to reach low speed	When the RG is performing the variable var1_out is activated	EXs combined with dlls

use software that allows reading memory locations. In this research, version 7.1.9 of INCA® was used [30].

- The automation process can be carried out in different ways: by using Python script or AutomationDesk® software [33]. In this research, the Python script was chosen because the staff's skill in AutomationDesk® in the service subjected to this case-study was low.
- Matlab® R2013 and Microsoft Visual Studio 2015 were used to create the dlls used in this research.
- Matelo®. Software used for validation purposes being able to generate test-cases.

2.2.3 *Results*

2.2.3.1 Ease for Automation Test-Cases

a. Simple software modules

Simple SMs, as indicated in the previous section, are characterized by handling a small number of variables. As a result, it is not difficult to reach the values established in the test-case. The problem associated with SM interactions appeared in all SMs considered in this research. For example, by analyzing the measurements obtained in the HIL simulation when validating a simple SM, by using MDA® [33], it was observed that, when actuating the brake pedal, multiple variables were affected and changed their values. When the brake pedal is actuated, the vehicle speed is reduced significantly, even without changing the accelerator pedal position. To decrease the vehicle speed, the engine ECU must control the engine combustion by modifying the air-diesel mixture rate. This phenomenon is regulated by other SMs which were not validated in this process. Therefore, one can conclude that to achieve the values set in the test-case, multiple SMs must be controlled simultaneously. This fact involves a great deal of complexity to code Python scripts.

One of the most important issues to be analyzed is the consequences of not reaching the values set in the test-case. Table 2.6 shows the results when validating the simple functions by using different techniques. As one can see, the tester-in-the-loop technique offers better results than the automated one without using dlls, because a technician makes the engine ECU reach a specific operating point during the test-case execution. When using dlls, the results are by 4.8% and 14.4% better than the tester-in-the-loop or automation results achieved by using a Python script only.

The Simulink® blocks that, in most cases, prevent reaching the values set in the test-case in this research, are show in Table 2.7.

Table 2.6 Comparisons of different techniques for validating simple SMs

Methodology	Number of cases in which the output value set in the test-case was no longer valid	Error rate after 250 simulations (%)	Success rate (%)
Automated with a Python script but without using a dll/model-based testing	49	19.6	80.4
Tester-in-the-loop	25	10	90
Automated with a Python script and the use of dll	13	5.2	94.8

Table 2.7 Most problematic Simulink blocks

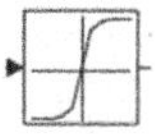	Interpolator block. In this case, depending on the input values presented to the Simulink® block, an output value is provided by applying an algorithm or an interpolation method
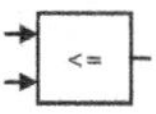	Simulink® native comparator block. It has problems in all its versions (greater than, greater than or equal to, less than, less than or equal to). In engine ECU software, on many occasions the value of a certain physical magnitude (e.g., motor revolutions, vehicle speed) is compared with a calibration threshold

It is important to analyze the root cause of the 5.2% failures. After the analysis of the 13 failures shown in Table 2.6, it was verified that the dynamic model used for the HIL simulation failed. Analysis showed that this issue came from 2 SMs. These SMs needed a 10 ms-sample period. Owing to imperfections of the HIL model, latency times and hardware limitations of the HIL bench, in certain occasions this sample time was not respected.

b. Fairly-complex and highly-complex software modules

For fairly-complex and highly complex validation SMs, the number of variables increased up to 80. Therefore, the issue of SM interactions is even more present. Figure 2.3 shows the total number and types of variables of a fairly-complex SM and the difficulty of manipulation to make the variables reach a specific value set in a test-case. The graph depicted in Fig. 2.3 shows that the Boolean variables were easier to be manipulated to reach the desired value, especially when they were related to variables directly linked to the driver's interface-model. If they were linked to analogical variables, it was not easy to reach the desired value. The triangle obtained for a fairly-complex SM was an isosceles whose height is focused on high difficulty. Therefore, the issue about SM interaction arises. On average, after having analyzed 5 SMs it was concluded that at least 40 variables were influenced between them. It is important to explain the nuance of "at least". The Boolean variables are simple to manipulate. Nevertheless, some of them have a direct impact on making the analogical ones reach the desired value established in the test-case. The HIL simulation results are shown in Table 2.8 in which one can see the number of times the expected output values specified in the test-cases are no longer valid when the SM inputs fail to reach the specific values set in the test-case. At the same time, the most problematic blocks present in the Simulink® models can also be observed (Table 2.9).

When it comes to a highly complex SM, the triangle obtained is closer to that of an isosceles one with a lower base. This characteristic indicates a greater presence of variables that are difficult to manipulate in a HIL simulation (Fig. 2.4). In this case, a total of 120 variables that influence the other variables had to be handled. The Simulink® blocks that pose the most problems were the same as those shown in Table 2.8. The results after the 100 HIL simulations are shown in Table 2.10.

In highly complex SMs, errors that prevent the HIL simulation from succeeding when using dlls were also detected. When validating a highly complex SM, a lower

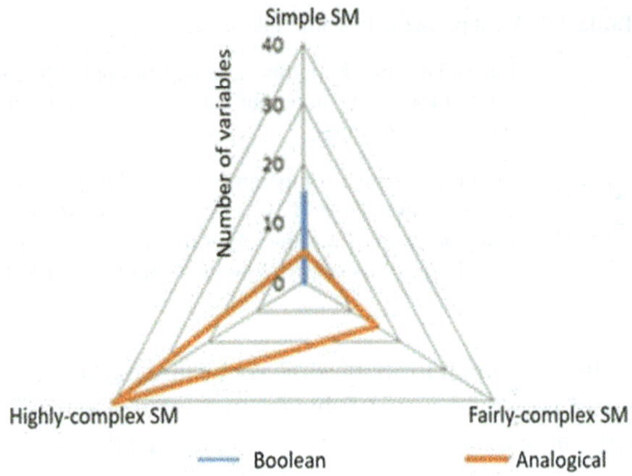

Fig. 2.3 Type of variables present in an average-complexity SM

Table 2.8 Comparisons of different techniques for validating fairly-complex SMs

Methodology	Number of cases in which the output value set in the test-case was no longer valid	Error rate after 1250 simulations (%)	Success rate (%)
Automated with a Python script but without using a dll/ model-based testing	480	38.4	61.6
Tester-in-the-loop	200	16	84
Automated with a Python script and the use of dll	125	10	90

success-rate with dlls was obtained because these SMs require covering thousands of kilometers (close to 20,000 km in some cases). Thus, the probability of failure in the simulator increases. Considering the strong SM interaction, it is unlikely to reach the specific values set in the test-case. Thus, the tester-in-the-loop solution offers worse results than when using dlls.

In fairly and highly complex SMs, at any given time, it was observed that several variables were close to the values previously set in the test-case as long as other values were quite far. If some manipulations were performed to make all the variables closer to the values set in the test-case, then the ones which were far from the expected values started to get closer, and the remaining variables started to get further. Thus, it is unlikely to be able to reach the input values set in a test case owing to SM interactions in such complex software as in an HIL simulation. Figure 2.5 shows how, by increasing the error tolerance against the value set in the test-case for the variables that constitute the test-case, the number of variables that remained within those tolerance margins increased. However, in any case, it was never possible to make

Table 2.9 Most problematic Simulink blocks for an average SM

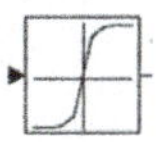	Interpolator block. In this case, depending on the input values presented in the Simulink® block, an output value is provided by applying an algorithm or interpolation method
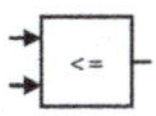	Simulink® native comparator block. It has problems in all its versions (greater than, greater than or equal to, less than, less than or equal to). In engine ECU software, on many occasions the value of a certain physical magnitude (e.g., motor revolutions, vehicle speed) is compared with a calibration threshold
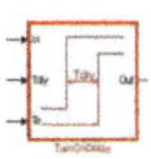	This block sets the output to TRUE while the input In remains TRUE for a certain calibratable time. Otherwise, the output is FALSE. As found in this research, when it comes to average and complex functions, it is more difficult than in simple functions to succeed by making the input In remain stable
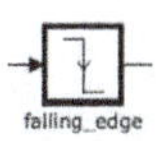	This block provides a Boolean type TRUE when a falling edge is detected. Otherwise, it remains FALSE. In this case, when it comes to average and complex functions, it is difficult in certain cases (for example when validating exhaust gas treatment systems) to reach the conditions to generate a falling edge
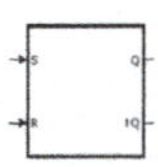	This block works as a typical RS flip-flop. As in a falling edge block, when it comes to average and complex functions, it is difficult in certain cases (for example when validating exhaust gas treatment systems or oil adaptive maintenance function) to reach the conditions when the S-input could be activated

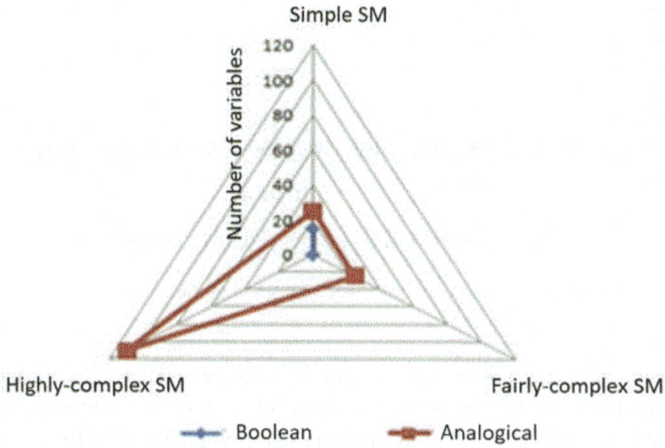

Fig. 2.4 Type of variables present in a high-complex SM

all the variables remain within the established tolerance range. This fact happened when executing the test-cases manually or automatically. As a result, these results show the great difficulty of validating an engine ECU software version by using HIL simulation.

Figure 2.6 summarizes the results obtained when using or not using dlls in an HIL simulation. As shown, dlls improve the HIL results in a significant way for all types of SMs, especially for simple and fairly-complex functions. It must be reminded

Table 2.10 Comparisons of different techniques for validating highly complex SMs

Methodology	Number of cases in which the output value set in the test-case was no longer valid	Error rate after 100 simulations (%)	Success rate (%)
Automated with a Python script but without using a dll/ model-based testing	61	61	39
Tester-in-the-loop	35	35	65
Automated with a Python script and the use of dll	15	15	85

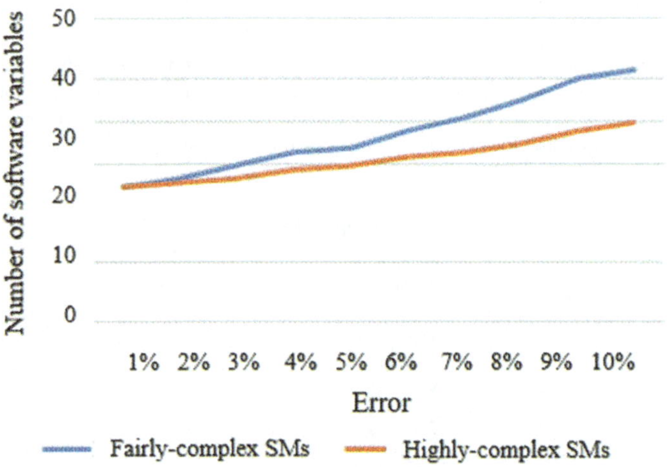

Fig. 2.5 Error trend depending on error tolerance of the SM inputs

that estimator SMs belong mainly to simple functions. That is why one can see such a huge difference when comparing the results obtained when activating or not activating dlls in an HIL simulation. In fairly and highly complex functions, it must also be noted that the SMs that require performing of many calculations belong to this category. Thus, there is also a significant difference when using dlls.

The reader may think that the automation process is not useful when validating the engine ECU software. This conclusion is false as there are some SMs, especially those related to electronics, which can be successfully automated such as CAN (Controller Area Network) and LIN (Local Interconnect network) bus or the basic functionalities of adaptive cruise control with the capacity to stop the vehicle (see Table 2.11). These statements have been proven in this research as shown in Table 2.12.

In this research, the SMs listed in Table 2.10 were used.

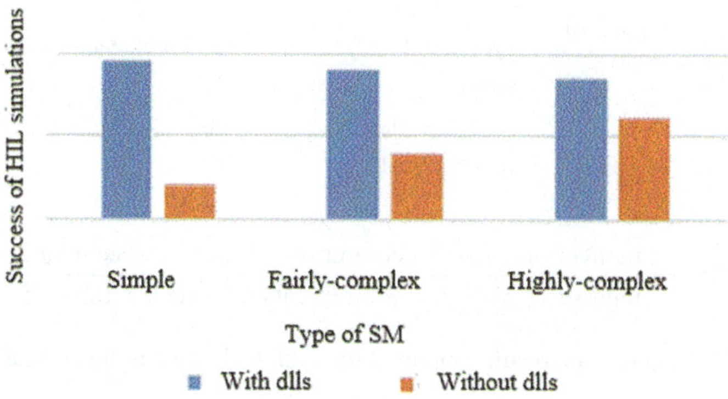

Fig. 2.6 Comparison of results obtained when using and not using dlls

Table 2.11 List of SMs used and tested in this research

Functions or SMs tested	Number of test-cases tested
CAN Bus	600
Driving aid systems	140
Pressure and Temperature carburant probe (SENT)	100
LIN Bus	50

Table 2.12 Comparisons of different techniques for validating functions depicted in Table 2.10

Methodology	Number of cases in which the output value set in the test-case was no longer valid	Error rate after 1000 simulations (%)	Success rate (%)
Automated with a Python script but without using a dll	12	1.2	98.8
Tester-in-the-loop	13	1.3	98.7

Table 2.11 does not show the results for automation with dlls as most of the function did not have a Simulink® model.

2.2.3.2 Functional Coverage

The functional coverage has been assessed by using Eq. (2.1) which is widely employed in the automotive sector. Table 2.13 shows the total number of functional requirements associated with the SMs validated in this research.

Table 2.13 Number of total functional requirements

Type of SM	Number of requirements
Simple	75
Fairly-complex	400
Highly-complex	510

$$\mathrm{FC} = \frac{\text{number of software requirements tested by a technique}}{\text{number of software requirements indicated in Table 15}} \quad (2.1)$$

Table 2.14 depicts the results obtained for each technique in this research.

a. Cause-effect technique and tester-in-the-loop

All test-cases run in this research by using these techniques are similar to the ones depicted in Table 2.5. It must be reminded that the test-cases can be run in a manual way or by employing Python scripts with the aim of automating the process.

The main limitation of the cause-effect technique is test-case redundancy. Many test cases run to validate the software were indeed linked to the same software requirements. The main reason behind this issue is the lack of a functional model of the SM under validation. When a use-case is not considered initially in the software requirements, it cannot be found by the cause-effect technique. In addition, bugs linked to calculation errors cannot be detected.

b. Model-based testing

When using Matelo®, it is important to expose the problems found. If the test engineer let Matelo® generate test-cases, this software will assign specific values for each input of the SM under validation. As a consequence, the problems of SM interactions, are identified. The only way to overcome this issue is to use fuzzy values combined with dlls. In this case, results are similar to the ones obtained when using a performance EX as long as dlls are used. Matelo® can be used also in such a way that Matelo® will not generate the test-case but it will control the automation process. In order words, the test engineer must code a Python script to generate the test -cases

Table 2.14 Functional coverage obtained for each research

Technique	Simple SM		Fairly-complex SM		Highly-complex SM	
	Number of rules tested	Functional coverage (%)	Number of rules tested	Functional coverage (%)	Number of rules tested	Functional coverage (%)
Cause-effect	64	85.3	312	78	357	70
Model-based testing	64	85.3	312	78	357	70
Tester-in-the-loop	64	85.3	312	78	357	70
Performance EX combined with dlls	68	90.7	348	87	445	87.2

needed and then Matelo® will check the functional states covered as the automation is performed.

In the present research, the test engineer codes Python scripts with the aim of running the same test-cases as for the manual execution, the tester-in-the-loop and so on. Consequently, the results shown in Table 2.9 are the same for the cause-effect technique and the model based-testing one.

c. Performance expert system

The rule-based EX allows specifying the functional requirements of SMs. Two phases are considered when validating EXs: a validation and a test one. On the one hand, the former consists of verifying a certain number of test-cases depending on the type of SMs to assess the EX performance to be sure that the EXs seem to work properly (Table 2.15). Table 2.16 shows the results obtained during the first phase in which a 83.3% success rate was obtained.

Once the errors were corrected, the test phase was performed to assure that the EXs would assess the software behavior properly. If no error occurred the EX was accepted.

The main conclusion that can be drawn is all possible use-cases are not checked when no EX is used. When it comes to simple and medium-complexity SMs, the number of unchecked functional states is shown in Table 2.17. The number of untested rules in a medium or highly complex function is greater because of the large number of use cases involved in this type of SMs.

Table 2.15 Number of test-cases used to validate the EXs

	Number of test-cases used to test the EX during the verification process	Number of test-cases used to test the EX during the acceptance process
Simple SMs	100	80
Fairly complex SMs	120	50
Highly complex SMs	5	2

Table 2.16 Errors detected when validating the EXs

Type of error	Percentage (%)	Cases	Explanation
Wrong syntaxes	8.8	20	Because the rules used to design the EXs are extremely complex, the programmer made coding errors
Incoherence between rules	3.5	8	In some cases of wrong performance of the EX, incoherence between rules was found
Misunderstanding of technical specifications	3.1	7	Because of innovative evolutions in some parts of the engine, some technical specifications were not understood properly
Rules not coded or forgotten	1.3	3	This error is owing to the same misunderstanding of technical specifications

Table 2.17 Number of rules or functional states not checked when an EX is not used

Type of SM	Number of functional states not tested without using an EX
Simple	4
Fairly-complex	36
Highly-complex	88

These improvements are mainly based on two reasons:

1. Dlls allow controlling better the HIL simulation as it is possible to know at any time if the current state of the engine ECU is coherent or not as already exposed in this research.
2. EXs assess the functional coverage easily. The reader can think that a similar result could be obtained by using Matelo® combined with dlls. Matelo® generates test-cases off-line. If after the HIL simulation, the inputs of SM under validation do not reach the desired operating point, Matelo® cannot calculate the expected value for the current state of the engine ECU in-real time.
3. Dlls allow finding bugs linked to calculation errors.

2.2.3.3 Productivity Gain

It is essential to check if EXs implementation respects the timeframes of the project by analyzing several factors. As shown in Table 2.18, the gain is positive for fairly and highly-complex SMs when using an EX. This gain comes from the automation process which allows testing test-cases quicker. In addition, these test-cases can be always run thanks to dlls. Consequently, an EX combined with dlls performs better than the other techniques. For simple SMs, the result is different as the HIL simulation implies that very simple and quick manipulations are conducted on the driver's interface model. As a result, the time gain is negative and the timeframe of the project may not be respected. It must be reminded that several projects are being developed at the same time by car manufacturers: diesel or gasoline engines. Between these types of engines, one can find considerable differences when it comes to torque structure or after treatment of exhaust gas systems. However, when comparing engines of the same groups, they are remarkably similar. As a result, an EX designed for a project can be used for another one. Then, only the automation and validation phases will be performed. As one can see in these phases, this technique outperforms the other ones. The main conclusions which can be drawn is that the proposed technique always meets the project planning especially when there are several engines developing at the same time.

Table 2.18 Time needed to design test-cases and rule-based EXs

		Simple functions	Fairly complex function	Complex function
	Total time for designing test-cases (h)	8	80	120
Time for designing and coding	Time for coding, design and validate EXs and Python script for the automation process (h)	4	35	70
	Time for preparing dlls (h)	2	6	10
	Time for coding a Python script (h)	4	32	50
	Time for coding when using the tester-in-the-loop (h)	2	25	35
	Total time for designing and coding when using EXs (h)	14	121	200
	Total time for designing and coding when using Python scripts (h)	12	112	170
	Total time for designing and coding when using the tester-in-the-loop (h)	10	105	155
	Total time for designing when executing a test-case manually (h)	8	80	120
Test-case execution	Time for executing an automated test-case by using EXs (h)	0.32	13	73
	Time for executing an automated test-case (h)	0.25	12.5	72
	Time for executing a test-case by using the tester-in-the-loop (h)	0.46	62	80
	Time for executing a test-case manually (h)	0.5	80	170
Validation	Time for validating the results with automation (h)[a]	0.00028	0.00347	0.00044
	Time for validating the results without automation (h)	1.67	20.83	2.33
Total time	Total time by using EXs (h)	14.32	134.00	273.00
	Total time with automation by using Python scripts (h)	12.25	124.50	242.00
	Total time when using the tester-in-the-loop (h)	10.46	167.00	235.00
	Total time without automation (h)	10.17	180.83	292.33

[a]In this case, the following data have been considered: 50 test-cases for simple functions with an execution time of 0.02 s, 250 test-cases for fairly complex functions with an execution time of 0.05 s, and 50 test-cases for complex functions with an execution time of 0.08 s. The execution time was measured by using the Python function time clock

2.2.3.4 Bug Detection

Figure 2.7 shows the bugs found by each technique when running the test-cases. The tester-in-the-loop offers a better performance than the automation process as it can make the system reach critical states that are not easy to reach when only using a Python script. There are not significant differences between manual and tester-in-the-loop techniques when it comes to bug detection as there is a technician who participates in the test-case execution, Python scripts detect fewer bugs than the rest of the techniques as test-cases are difficult to automate due to SM interactions. As a consequence, when the system reaches an operating point close to the one established in the test-cases, the outputs indicated in the test-cases may be no longer valid. To solve these problems, fuzzy values for the SM inputs may be used as exposed later in this section.

The results obtained in this research show that EXs with dlls give better performance and can be used to test more functional states and detect more bugs than the other techniques. Basically, this statement is based on two main reasons:

1. The problems coming from the SM interactions are fixed due to dlls. Even though the operating point established in the test-case is not reached, dlls can provide the right values expected from the software. Consequently, the test-cases can be successfully run and the automation process can validate the HIL simulation results automatically.
2. The functional coverage is improved due to the existence of the functional model. In addition, this model can be covered easily thanks to the automation success by using the dlls. It is also important to establish the main types of bugs found for each technique (Table 2.19).
3. When the bug is linked to calculation errors (calculation faults).

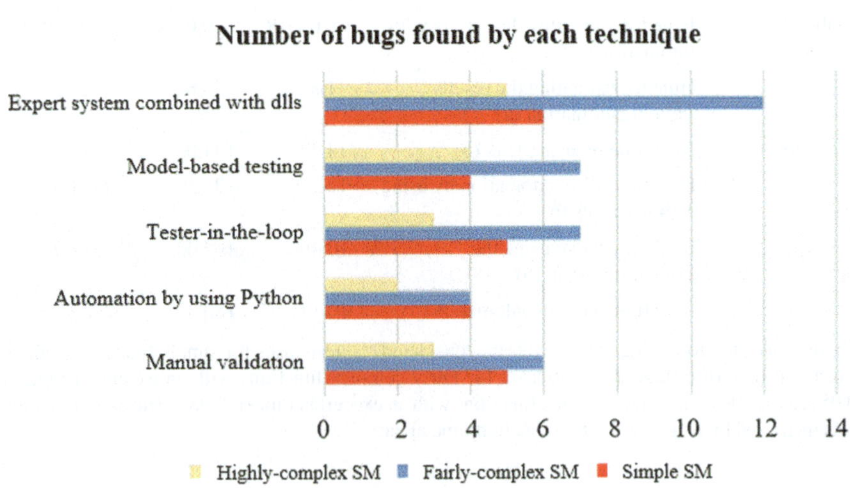

Fig. 2.7 Bugs found when using different techniques

Table 2.19 Type of bugs detected

	Calculation faults	Bugs	Performance faults
Manual validation	0	12	2
Tester-in-the-loop	0	14	1
Automation without dlls	0	10	0
Model-based testing	0	10	5
EXs and dlls	5	14	4

4. When no code error occurred but there was unexpected performance software. This issue can come from an error design in the SM under validation (performance faults).
5. When there is a code bug. This means the programmer has made a mistake and coded differently from what was indicated in the specifications.

2.2.3.5 Costs

It is necessary to discuss costs. The first one is associated with the licenses needed to use a specific technique (already discussed). The other one is linked to software versions needed to correct bugs detected at the end of the project. This can be caused by two things. Firstly, certain SMs (especially those related to advanced driver assistance systems) cannot be tested at the beginning of the project. The validation of these functions needs very mature software of some ECUs present in the network (electronic stability program ECU, body control unit, radars, cameras, gearbox ECU in automatic cars). Secondly, some bugs appear when testing some use-cases that were not considered in the validation process. When these bugs are detected, the project team must decide whether the bug has a significant functional impact and therefore require correction of the software. Otherwise, the bug can be corrected in future engine projects and no correction will be made. Developing new software versions involves a high cost but also might imply updating the ECU of vehicles that have already been marketed. The results showed that EXs combined with dlls detected two bugs that would have required corrective software development. These bugs were not detected using the cause-effect technique, the model-based testing one, the manual execution or the model-based testing one.

The reader might think that, in case of bugs in the Simulink® model, the software will also contain these faults. As a result, no bug will be detected by using the method proposed in this research. This study has proven that this statement is true and that is why the performance EX must be used.

2.2.3.6 Comparison Among Other Methods

When performing an HIL simulation, it is not easy to reach the values indicated in the test-case due to SM interactions. Figure 2.8 shows an example of a histogram

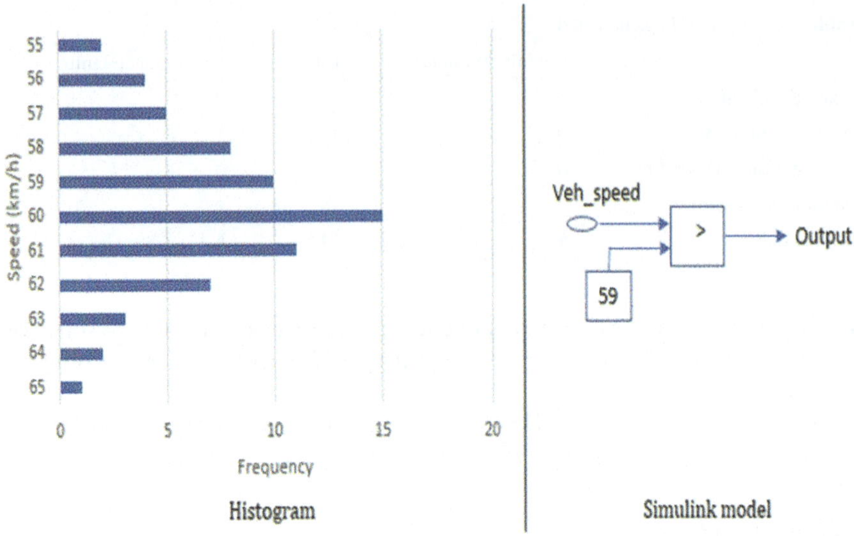

Fig. 2.8 Example of test-case

displaying speed value. Depending on the value reached, the output can be 1 or 0. Consequently, if a test-case indicates that the speed must be 60 km/h, the accuracy is a critical factor and the expected output could be no longer valid.

A comparison among different techniques is shown in Table 2.20.

2.2.4 Conclusions

This research, conducted at the second most important European car manufacturer, is focused on the software validation of an engine ECU by using dlls and an EX (ES). This combination allows the detection of software performance and coding bugs. As shown in this research, dlls and ES can detect bugs that other techniques such as the black-box or the tester-in-the-loop cannot, especially those in temperature estimator SMs and after-treatment of exhaust gases SMs, which require accurate calculations. The obtained results show how dlls and the EX can improve the HIL success rate compared with the tester-in-the-loop technique and can execute 4.8% of the test-cases in simple validation SMs, 6% of the test-cases of fairly complex SMs and 20% of the test-cases of highly complex SMs despite the presence of SM interactions. In comparison to the use of a Python script without using a dll, the dlls and the EX can improve the HIL and can execute 14.4% of the test-cases in simple validation SMs, 28.4% of the test-cases of fairly complex SMs and 46% of the test-cases of highly complex SMs. As a result, dlls can overcome the issue linked to SM interactions. In addition, between 9 and 13 more bugs were found when using the EX and dlls, six of

Table 2.20 Comparison among different techniques

<table>
<tr><th></th><th>Manual validation</th><th>Automation without dll</th><th>Model-based testing</th><th>EXs and dlls</th></tr>
<tr><td>Validity of test-cases</td><td colspan="3">As shown in Fig. 2.9, it is necessary to reach the exact value indicated in the test-case. Otherwise, the validation process cannot be performed automatically</td><td>Even though the values indicated in the test-case are not reached, the validation can be performed automatically</td></tr>
<tr><td>Accuracy needed</td><td colspan="3">The test-case output may be no longer valid (Fig. 2.9). The test-engineer should check the specification to confirm the expected output</td><td>The test-case output may be no longer valid. However, dlls can check the expected output automatically</td></tr>
<tr><td>Complexity</td><td colspan="3">As shown in Fig. 2.9, it is highly complicated to reach the specific values indicated in the test-cases (see Tables 2.6, 2.8 and 2.10) due to SM interactions</td><td>Even though the HIL simulation does not reach the specific values indicated in the test-case, the validation process can be performed</td></tr>
<tr><td>Robustness in case of failure</td><td colspan="3">During the HIL simulation, the engine ECU can detect a failure (low rail pressure, turbo failure, etc.). In that case, the test-case output is no longer valid</td><td>Even though the engine ECU detects a failure, the dll can detect the expected output in that case</td></tr>
<tr><td>Reading ECU variables in real-time</td><td colspan="3">INCA software does not allow reading in-real time variables by using Python while data acquisition is performed. The test-engineer has to analyze the data acquisition to check if a bug is present</td><td>The dll can do the validation process automatically when using a C-code at the same time</td></tr>
</table>

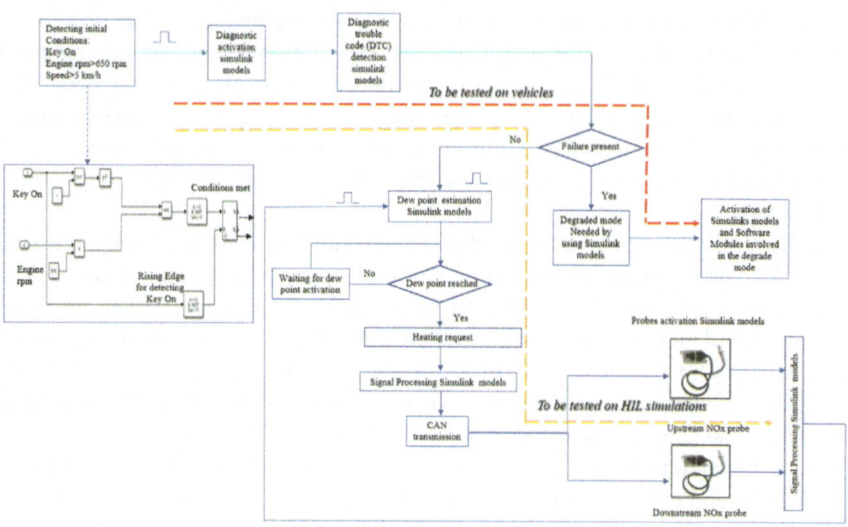

Fig. 2.9 Example of model and activation conditions

which could not be detected by other techniques. Even though EXs and dlls require more time to be implemented, the timeframe of the project was respected.

2.3 Use of Genetic Algorithms to Reduce Costs of the Software Validation Process[2]

2.3.1 Introduction

The number of ECUs installed in vehicles is increasingly high. Manufacturers must improve the software quality. Innovative techniques must be proposed to reduce cost and increase software quality.

This research proposes a technique being able to generate not only test-cases in real time but to decide the best means to run them (HIL simulations or prototype vehicles) to reduce the cost and software testing time. It is focused on the engine ECU software which is one of the most complex software installed in vehicles. This software is coded by using Simulink® models. Two genetic algorithms (GAs) were coded. The first one is in charge of choosing which parts of the Simulink® models should be validated by using HIL simulations and which ones by using prototype vehicles. The second one tunes the inputs of the SM under validation to cover these parts of the Simulink® models. The usage of dlls is described to deal with the issues linked to SM interactions when running HIL simulations.

GAs found at least 7 more bugs than traditional techniques and improved the functional and code coverage by between 3% and 11% for functional coverage and by between 1.4% and 7% for code coverage depending on the SM complexity. The validation time is reduced by 11.9% regarding traditional techniques. GAs perform better than traditional techniques improving software quality and reducing costs and validation time. The usage of dlls allows testing the software in real time as described in this study.

Both the number of ECUs installed in vehicles and their complexity are increasing [5, 34, 35]. Thus, manufacturers must assure software quality and reliability [12]. The software and hardware validation of an engine ECU is performed by using the HIL simulation and prototype vehicles [36]. The HIL simulation has several advantages as no vehicle with all ECUs updated with the latest software version is necessary. Secondly, the ECU behavior in the network can be checked by analyzing the frames transmitted and received when conducting an HIL simulation. However, the real interactions between ECUs are not tested as all frames received are sent by a model and not by real ECUs. Regarding prototype vehicles, the engine ECU software is tested in real vehicles which must have all ECUs properly updated: ESP

[2] Extracted from Ortega-Cabezas, P.M., Colmenar-Santos, A., Borge-Diez, D. et al. Experience report on the application of genetic algorithms to reduce costs of the software validation process in the automotive sector during an engine control unit project. *Software Quality Journal* 30, 687–728 (2022). https://doi.org/10.1007/s11219-021-09582-x. https://www.springer.com/journal/11219.

(Electronic Stability Program), ADAS (Advanced Driver-Assistance System ECU), ATCU (Automatic Transmission Control Unit), etc.

This chapter is focused on one of the most complex software installed in vehicles: the engine ECU software. It proposes the usage of GAs aiming at choosing the most adequate means to be used for validation while generating test-cases automatically at the same time. The main goals are:

a. Choosing automatically the optimal means to reduce the validation time and costs.
b. Finding solutions to technical problems when using the HIL simulation due to SM interactions.
c. Assessing whether GAs perform better than other techniques such as the model-based testing and the black-box techniques.
d. Verifying whether GAs are able to find bugs when other techniques fail.
e. Assessing the staff skill impact on the validation process.

The engine ECU software development comprises three phases (V-cycle development): implementing models based on Simulink® software in order to control the engine performance, generating C-code and checking the final integration of the software into the hardware. During the whole process, the engine software completes three levels of testing: model-in-the-loop (MIL), software-in-the-loop (SIL) and HIL simulations [37]. Consequently, the software is tested to assure that it meets all requirements. During the MIL, a controller model is implemented and applied to the Simulink® model aiming at checking if the model behaves as expected [38–40]. During the HIL simulation, the integration between software and hardware is tested thanks to a controller (the engine ECU and its software) which controls the system that imitates the engine behavior (the HIL simulator) [41–45]. In addition, prototype vehicles are used to test some functions which cannot be completely validated when using HIL simulations such as ADAS [46]. Therefore, the most adequate means to validate the software must be chosen to reduce time and costs. Finally, SIL is employed to test an executable code within a modelling environment [47].

Currently, software is tested based on software, architecture and system requirements [37]. At this point, how to test software requirements is a key point discussed in some standards such as ASPICE (2020). Software testability depends on 5 factors such as: requirements, built-in test capabilities, the test-cases design, the test support environment, and the software process in which testing is conducted [48]. Regarding software requirements, the most significant cause of accidents due to software is linked to poorly created software requirements or requirements that are partially delivered to developers [49, 50]. Dos Santos et al. carried out a detailed analysis about software requirements testing approaches such as the requirement driven testing [51, 52].

Concerning autonomous driving, ISO 26262 only covers functional safety when a failure occurs but not when there is no system failure. That is why, the safety of the intended functionality (SOTIF) ISO 21448 came out [53, 54]. Some key topics to validate the software are focused on 3D Modeling and sensor buildings. The former aims to create a realistic environment while the latter consists of modeling and testing sensors among others [55]. Huang et al. detail in their research the main tendencies to validate software such as software testing, simulation testing, x-in-the-loop testing and driving test in real conditions [56]. Riedmaier et al. describe an important method to test the software: the scenario-based approach which allows individual traffic situations to be tested by using virtual simulations [57]. Other approaches such as formal verification, a function-based approach, real-world testing, shadow mode testing and traffic-simulation-based approach are used to test SOTIF. The main difference among them is that in the scenario-based and function-based approaches, a microscopic statement about the safety of the system is first made to be transferred to a macroscopic statement. The rest of the methods result directly in a macroscopic statement. There are solutions in the market which allow rapid prototype, MIL/SIL simulations, HIL simulations and real test drives [58].

Cybersecurity in the automotive industry involves three main factors to be considered such as authentication and access control, protection from external attacks and detection and incident response [59]. The factors which make the automotive security more efficient include integration of right solutions such as firewalls, protecting communications, authenticating communications and encrypting data [60, 61]. These topics are important to offer performance such as on-the-air software update and V2X communication [62, 63]. As detailed by McAfee, the scope of cybersecurity involves the distributed security architecture, hardware and software security and finally network security [62]. Standards such as ISO/SAE 21434 will help the automotive sector to implement solutions for effective compliance with cybersecurity requirements as today's knowledge sharing is inadequate [64, 63] In this research, some topics linked to cybersecurity testing are analyzed.

2.3.2 *Methods*

2.3.2.1 Simulink Models

The SMs are composed of multiple complex Simulink® models and subsystems. Figure 2.9 shows an example of the internal structure of the SM linked to the NO_x heating probes installed in vehicles. When the initial conditions are reached (key on, the engine rpm more than 650 rpm and the vehicle speed higher than a threshold), the engine ECU software checks whether the dew point is reached. This point is the temperature to which air must be cooled to become saturated with water vapor. Afterwards, the NO_x probes start to heat until reaching the required temperature to measure NO_x ppm present in the exhaust gas pipe. In this study, all models were transformed into models based on nodes (S_x) which represent different low-level system states[3] and relations between them (Fig. 2.10). When designing test-cases, it must be determined which parts of the Simulink® model should be validated by

[3] Low system states are functional states at low level. Consequently, the functional state cannot be detected by the driver.

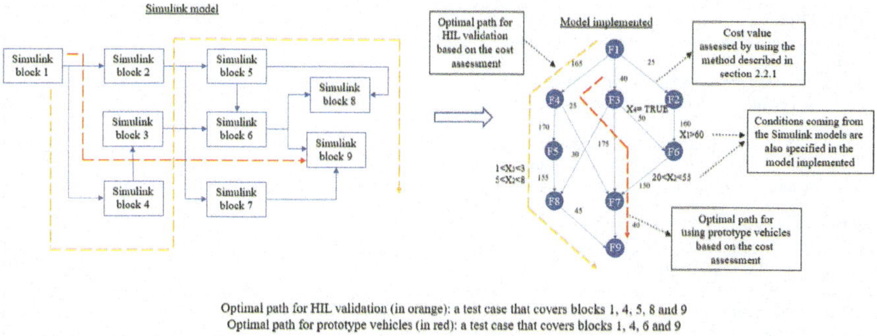

Fig. 2.10 Example of model and activation conditions

using the HIL simulation or prototype vehicles and how inputs are tuned. Next section describes in-detail how GAs work to do this.

2.3.2.2 How GAs Work Together

Figure 2.11 depicts a pseudocode and a high-level description of the method. A model is implemented by using the Python code (Fig. 2.11) through the variable named *ARCS* which contains the cost and conditions to go from one state to another one. Next sections display how the conditions are specified. Once the model is made, the GAs are parametrized, and the range values of the input variables of the SM under validation and the constrained linked to the optimization problem are specified. The GA2 generates populations (inputs for the SM under validation). GA1 is used to assess and optimize the path with the lowest cost by doing operations such as mutation or crossover taking into account the population generated by GA2. The GA2 makes the population evolve in such a way that the cost calculated by GA1 is minimized.

2.3.2.3 GAs Description

a. GA in charge of tuning inputs

Once the model is implemented and set in the code, this generates populations. To do this, the range must be specified as well as the constrained among software variables linked to the optimization problem. The fitness function of this GA2 is the output of GA1 described later which was responsible for finding an optimal path given specific inputs. When the GAs are run, the results display the values of all inputs of the SM which cover the path requiring the minimum cost and the usage of HIL simulations.

```
IMPORT packages
#Describes nodes and arcs. Here is the structure
to define them
# Si-Sk describes the transition for going to Si to Sk:
cost and conditions. #The variable range is also
specified.
ARCS= {"Si-Sk": {"cost": 150, "conditions":
{"variable x":[range],
call_assessment_conditions}}

#Define input ranges
RANGES= {"variable1": [0,10],
         "variable2": [0, 20]
        }
#Define GAs parameters.
GA1_PARAMS,
    "initial_state"=S1
    "final_state"= S9 #considering that there are 9
states
    "mutation_factor"= 0.7
    "gens"= 9
    "retain_factor"= 0.4

#Define GAs parameters
GA2_PARAMS
    "mutation_factor"= 0.8
    "gens"= 50      # number of members of a
population
    "retain_factor"= 0.15

DEFINE conditions among variables
#Generate initial population of input variables.
This function returns the #values of each variable.

#Example value: {variable1=2, variable2=2.7},
{variable1=2.6, variable2=3.5},

variable1,
variable2...=input_population_generate()

# For each member, check the path obtained
depending on conditions indicated #in ARCS
FOR j=1 to gens_GA2
  FOR i=1 to gens_GA1
    #Calling the first GA to find the best path. In
other
    #words,the path
    #with minimum cost for a given input
    cost, states, fitness_score =
    calling_GA1_find_best_path(inputs_values[i])
  NEXT
  evolve_mutation_retain(inputs)
NEXT
```

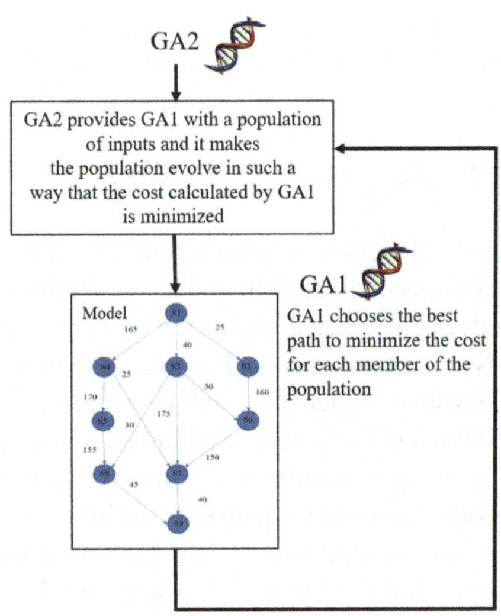

Fig. 2.11 Pseudocode of how GAs work together

b. GA in charge of choosing the most adequate means

Several key factors must be considered when deciding the most adequate means to validate the software such as the ones shown in Table 2.21 [42, 65].

All factors shown in Table 2.21 are assessed when going among states of the models (see Fig. 2.10). This process is composed of two phases:

Table 2.21 Factors considered to assess the fitness function

Factor	Description
Tuning activities	Some SMs must be tuned before its validation such as combustion/injection SMs. In this case, the engine software can apply different cartographies to inject the optimal amount of diesel or gasoline. If one of them is not tuned, the engine may stall. Consequently, a dataset which guarantees a minimum functionality of the SM under validation must be available
Time needed to go from one state to another one	An estimate of the time needed to validate an SM when using a vehicle or an HIL model is made. There are two possibilities—either to perform the simulation by using an HIL model or a vehicle. The former implies that the HIL model must be robust. The latter implies that a vehicle should be used. Some use-cases are difficult to reach when using prototype vehicles. Test-engineers' experience is essential to assess properly this factor
Dependency on ECUs	When validating a certain SM by using a vehicle, all ECUs must be updated aiming at assuring that all frames are properly transmitted and received among other factors. Otherwise, the validation process such as the adaptive cruise control SM cannot be performed. In this case, at least the ADAS, ESP and engine ECU must be properly updated and tuned
Risk level	The automotive safety integrity level is a system which classifies potential risks posed in the vehicle when it is operated by using the ISO 26262. For this purpose, it uses three parameters such as: exposure, controllability and severity with the aim of establishing a score. By using this score, a series of automotive safety integrity level is established. Regarding the engine ECU, the software must guarantee the passengers' and vehicle safety in a dangerous situation. Depending on the ASIL values (A, B, C, D) the level of risk will be different
Feedback from other projects	It is common that several engine projects take place at the same time. Consequently, feedback from other projects is of paramount importance. Therefore, when a bug is found in a specific engine software version, it is immediately communicated to other project teams so that they can check if there is a bug in some other engine software applications. Meanwhile client complaints are also considered in such a way that if a project receives a client complaint, it is transmitted to other projects, which could galvanize all necessary actions

- *Phase 1.* A multidisciplinary team assesses these factors aiming at determining the cost of each path by using the process depicted in Fig. 2.12. As a result, a model with the whole cost set for each transition is obtained (Fig. 2.10).
- *Phase 2.* This GA chooses the most adequate means to validate the SM by assessing the cost function given by Eq. (2.2):

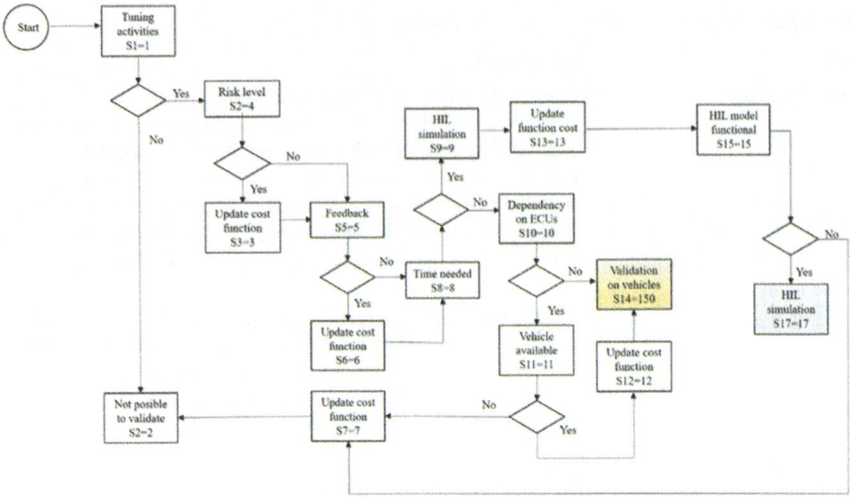

Fig. 2.12 Factors indicated in Eq. (2.1)

$$\text{Fitness function} = \sum_{i=1}^{i=n} S_i \tag{2.2}$$

where S_i is the cost of reaching the S_i state, $\sum_{i=1}^{i=n} S_i$ is the cost linked to all transitions of a specific path. When the HIL is chosen, the fitness function is always lower than 150. Otherwise, prototype vehicles are employed as, in this case, the fitness function reaches 150 or more. The reader can check this by adding all S_i values needed to reach S_{14}.

Each path is composed of different states. The paths which contain state 17 more frequently are considered as the optimal ones to be validated by using an HIL simulation. Otherwise, it should be validated by using prototype vehicles. The test-engineer can collect important information when analyzing the states covered once the optimal path is assessed (dependency on other ECUs, feedback from other projects, etc.).

2.3.2.4 HIL Simulations

Once the GAs are parametrized and a model is built as shown in Fig. 2.10, the HIL simulation can be conducted. In addition to the cost value, the actions to be conducted on the HIL model for each transition must be coded (Fig. 2.13) as the software variables have to reach the values specified in the test-case. Several ways to set the conditions to pass from one state to another one can be used. The first entails writing the equations directly in the code, which is limited to simple SMs as fairly complex and high complex SMs involve many equations. The second option is to call the

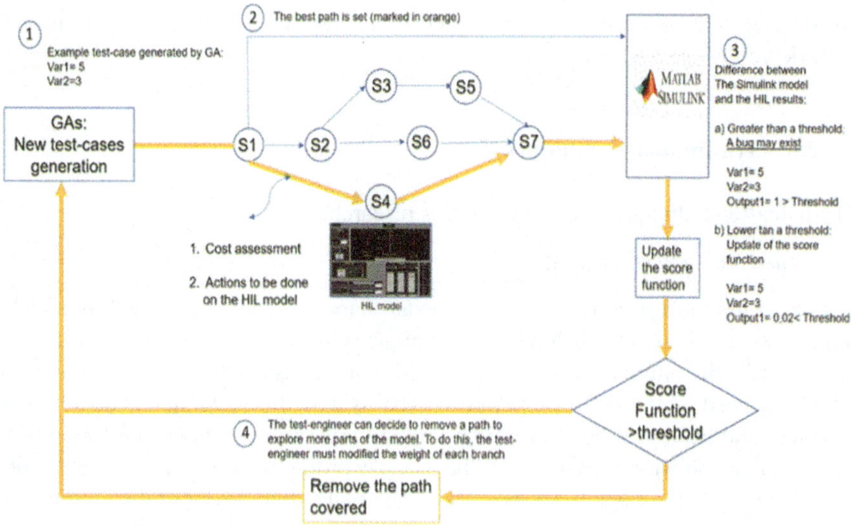

Fig. 2.13 HIL simulation process

Simulink® model by using the test-case inputs to make the Simulink® model return the expected output values. In this study, the Simulink® models were transformed into dlls by following the steps described in the official Matlab® documentation. Figure 2.13 depicts the usage of dlls. They are necessary to conduct the validation process to find bugs due to SM interactions as it will be shown in the results and discussion sections.

2.3.2.5 Network and Software and Hardware Integration

This proposal validates the network and hardware and software integration by using the dlls as shown in Fig. 2.13. Once the software is coded, the software outputs must be equal or very close (if the outputs are analogical) to the values provided by the Simulink® models despite the SM interactions. This point is checked by using the dlls which allow comparing the HIL results when running a test-case with the outputs provided by dlls. The same explanation can be used for vehicles as the data acquisition can be injected into the Simulink® models, and both results can be compared.

Regarding the network, it is tested when using prototype vehicles in real conditions. Not all SMs implemented in the software exchange information with other ECUs. All these aspects are considered in Fig. 2.12 where the reader can find state S_{10} which assesses if the SM under validation has an impact on the network. Anyway, if an SM must be validated and prototype vehicles with all ECUs properly updated are not available (specially at the beginning of the project), HIL simulations are used considering that the frames are simulated by using a model (this situation is also considered in Fig. 2.12).

2.3.2.6 Traditional Techniques

The hereafter techniques were used in this research.

a. The cause-effect technique

One of the most used techniques in the automotive sector is the black-box technique [18]. The main idea behind this technique is to test software as a black box. In other words, the internal structure of the SM is not considered by the test-engineer who is focused on the software behavior. That is why this technique is also known as behavioral testing. When designing the test to be run, test engineers design test-cases and decide which means could be used according to their experience [18, 66, 67]. The cause-effect is a black-box technique widely employed in the automotive sector for several reasons (easy to automate among others). This technique is based on considering a series of conditions linked to inputs of the SM under validation, the test-engineer must check if the software runs as expected. To do this, the test-engineer performs a series of actions by using the means employed for validation (prototype vehicles or the HIL simulation) and, finally, verify the software behavior. This behavior is validated and assessed by using the outputs of the SM under validation. It must be reminded that the means used to validate it are chosen by considering the test-engineers' experience when using this technique.

b. The model-based testing

It is a software testing technique consisting of deriving test-cases from a functional model which describes the functional aspects and requirements of the SM under validation. Thanks to this model, it is easier to assess the functional coverage as the number of functional states covered when validating an SM is known. When implementing it, all functional states and the transition from one state to another are indicated. In this research, Matelo® software was used to generate the functional model of SMs [68]. This software allows implementing a model easily. Regarding the activation of each transition, the conditions are set. In this study, each transition calls a Simulink® model to check the next state to be activated. Matelo® allows generating test-cases by assigning values to all variables used in transition in such a way that it tries to cover all possible transitions and paths. Finally, each state can be a model as it is the case in this research making the models extremely complex. Figure 2.14 sums up all aforementioned process explained. A test-case is generated and by using calls to Simulink® models, Matelo® determines which part of the model will be covered (Fig. 2.14 in orange). Many test-cases are generated to cover the whole model and to increase functional and code coverage.

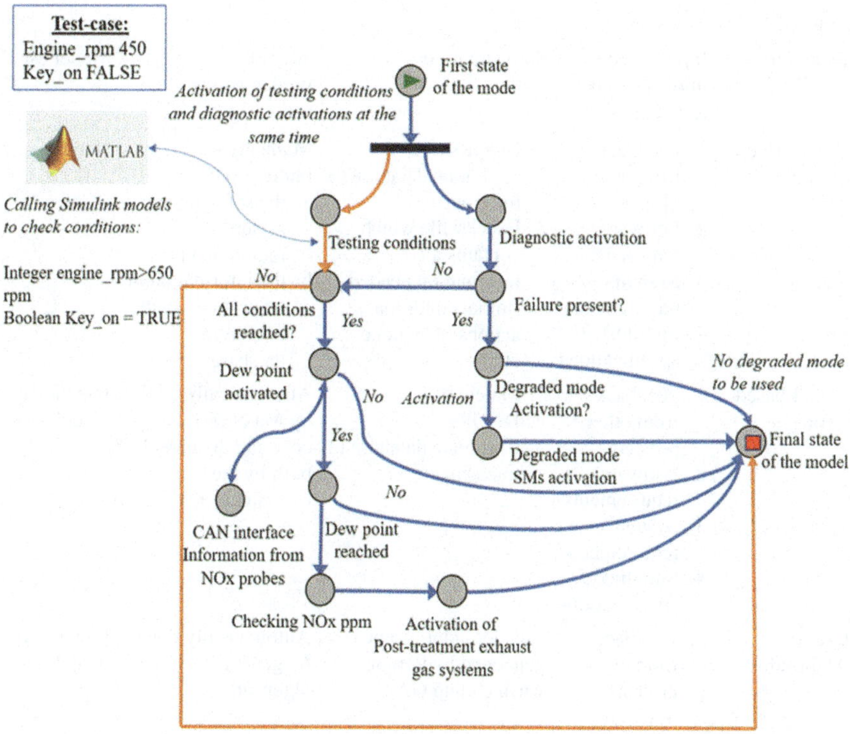

Fig. 2.14 Example of NO_x activation model based on Matelo®

2.3.2.7 Experimental Settings

The characteristics of SMs have an impact on three factors: the time needed to validate the software, the means used to run test-cases and the number of test-cases to be run considering the planning of the engine software development. According to the test-engineers' experience and the technical documentation used for coding the software, the SMs were classified as simple, fairly complex and high complex SMs (Table 2.2).

Table 2.22 shows the way of generating test-cases. All techniques used the software and system requirements traced in DOORs, feedback from other projects[4] and the Simulink® specifications as inputs. By analyzing all these input data, the test-engineers build models when using GAs and the model-based testing. Finally, test-cases are implemented automatically or manually. As described later, the test-engineers' skills have a significant impact on the time needed to implement test-cases and to obtain a productivity gain.

[4] Feedback from other projects means bugs found in a project which could impact another project.

Table 2.22 Test-cases run in this research

Technique	Inputs used for implementing test-cases	Software used	Way of implementing test-cases	Model used
Cause-effect technique	1. Feedback from other projects 2. Software requirements 3. System requirements 4. Simulink® specifications	1. DOORs 2. Corporate database to trace bugs 3. Excel® file which contains all information needed (initial conditions, actions to be done, etc.)	Manually by interpreting: a. the software and system requirements b. the information of bugs traced in the corporate database	None
Model-based testing	1. Feedback from other projects 2. Software requirements 3. System requirements 4. Simulink® specifications	1. Matelo® 2. DOORs 3. Corporate database to trace bugs	Automatically done by Matelo® by covering the model built by the test-engineer	Functional model
Genetic Algorithms	1. Feedback from other projects 2. Software requirements 3. System requirements 4. Simulink® specifications	Pseudorandom values generated by Python when coding GAs	Automatically done by genetic algorithms	Low level model

2.3.3 *Results*

This section compares the performance among GAs and traditional techniques by using the KPIs indicated in Table 2.23.

2.3.3.1 Code Coverage

During HIL simulations, a bug is detected if the difference between the HIL results and the outputs provided by the Simulink® models does not obey Eq. (2.3).

$$\sum_{j=1}^{j=m} \left|\mathrm{HIL}_j - \mathrm{Simulink}_j\right| \approx 0 \tag{2.3}$$

Table 2.23 KPI employed in this research

KPI	Description
Code coverage	It determines the number of Simulink® blocks successfully validated when running test-cases divided by the total number of Simulink® blocks considered
Functional coverage	It determines the number of functional states successfully tested when running test-cases divided by the total number of functional states considered
Validation software time	It describes the time needed to implement, run and validate an SM when running test-cases
Productivity gain	The time gain obtained when using a specific software validation technique
Bugs found and their types	Number of bugs and types found when using a specific software validation technique
Bugs found by other clients	Number of bugs found by other users of the engine ECU software such as ESP and ADAS validation staff

where HIL_j is the value for the output j of the SM under validation after having run a test-case by using an HIL simulation and Simulink_j is the value for the output j of the SM under validation after having run a test-case by using the Simulink® model.

The coverage is assessed by using Eq. (2.4) which relates to the number of Simulink® blocks tested versus the total number of blocks presented in the specifications of the SM under validation.

$$\text{Code coverage} = \frac{\text{number of Simulink}^{\circledR}\text{ blocks tested}}{\text{number of Simulink}^{\circledR}\text{ blocks present in the SM under validation}} \times 100 \tag{2.4}$$

Table 2.24 shows the number of blocks present in the SMs validated in this research, which is used to assess the code coverage (Table 2.25).

As the cause-effect technique does not use models, the code coverage is lower than the one obtained when using the model-based testing and GAs. Building a model in which each state is a Simulink® block allows testing the same functional state by following different branches of the Simulink® model (Fig. 2.15). The model-based testing does not allow tuning the inputs of the SM with the aim of choosing the best means (an HIL simulation or vehicles) to validate an SM contrary to GAs. In addition, this technique needs to define test-cases as inputs and expected outputs. In case of a problem with the automation process due to SM interactions, the expected outputs could be no longer valid. This problem is solved by GAs and dlls. Regarding GAs, the code coverage is the addition of the code coverage when using HIL simulation and

Table 2.24 Number of total Simulink® blocks

Type of SM	Number of Simulink® blocks
Simple	80
Fairly complex	350
High complex	530

Table 2.25 Code coverage obtained when validating the 15 SMs

Technique	Simple SM		Fairly complex SM		High complex SM	
	Number of Simulink® blocks	Code coverage (%)	Number of Simulink® blocks	Code coverage (%)	Number of Simulink® blocks	Code coverage (%)
Cause-effect	63	78.7	265	75.7	380	71.7
Model-based testing	68	85	285	81.4	410	77.3
GAs when using an HIL simulation	58	92.5	235	88.5	412	78.7
GAs when using prototype vehicles	16		75		5	

prototype vehicles. GAs perform better as they can cover more Simulink® blocks providing that the right means are used.

Code coverage should be at least 90% to meet standards. The validation process of an engine ECU is the combination of the software validation performed by the validation team (topic considered in this research), the tuning activities and the driving tests which consist of making 6 vehicles cover 20,000 km each to test the software in real conditions. The total code and functional coverage are assessed considering these three activities. No technique can reach 100% coverage due to several reasons such as project planning constraints. As proved later, validating by choosing the wrong means increases the validation time.

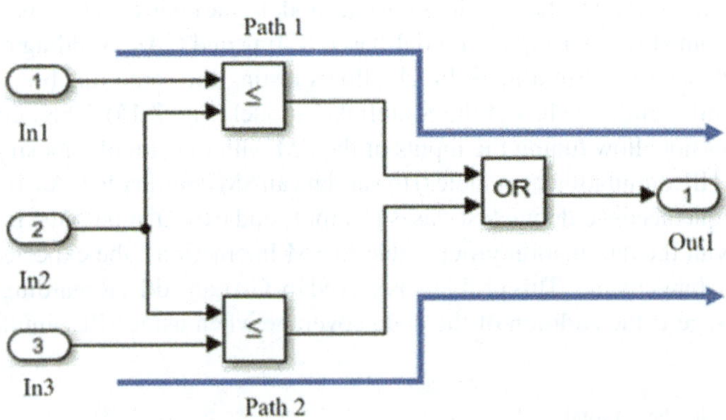

Fig. 2.15 Example of different ways of activating an output

Table 2.26 Number of total functional requirements

Type of SM	Number of requirements	Number of Simulink® blocks
Simple	75	80
Fairly complex	400	350
High complex	510	530

Table 2.27 Functional coverage obtained for each research

Technique	Simple SM		Fairly complex SM		High complex SM	
	Number of requirements tested	Functional coverage (%)	Number of requirements tested	Functional coverage (%)	Number of requirements tested	Functional coverage (%)
Cause-effect	60	80	302	75.5	357	70
Model-based testing	65	86.6	330	82.5	385	75.4
GAs	69	92	346	86.5	400	78.4

2.3.3.2 Functional Coverage

Table 2.26 shows the functional states linked to the Simulink® blocks present in the SM chosen. The number of functional states can be lower than the number of Simulink® blocks as some outputs of the SM can be activated by using several paths without any impacts on the functional state of the vehicle (Fig. 2.15).

Table 2.27 shows the results obtained for each technique. These results are logical as the higher the code coverage is, the higher the functional coverage is. The standard percentage of validation (90%) is reached thanks to tuning, validation and test-driving activities.

2.3.3.3 Automation

For several reasons, the automation process is difficult to be performed when it comes to engine ECU software due to SM interactions. Firstly, reaching the values for inputs of the SM under validation is difficult as the complexity of the SM increases. Secondly, if inputs do not reach the expected values, the values of the outputs of the SMs under validation will be no longer valid (Fig. 2.16).

Test-cases can be fully automated, partially automated or can be run manually. In this research, GAs were run by using the tester-in-the-loop and fully automated options. The success rate of reaching the values indicated in the test-case is shown in Fig. 2.17.

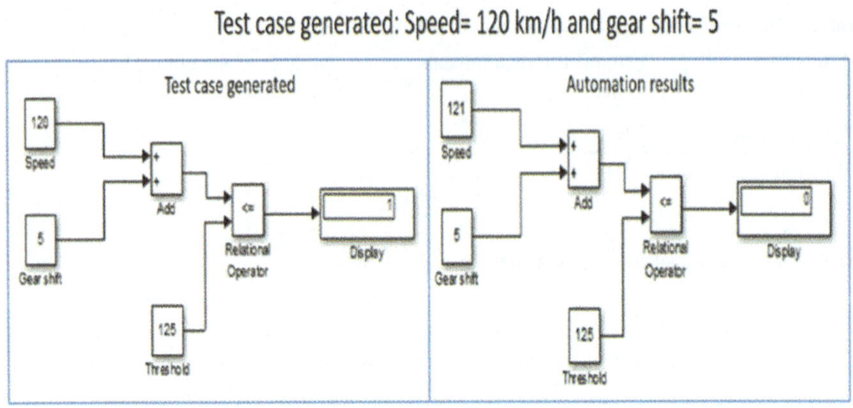

Fig. 2.16 Potential error when a test-case is automated

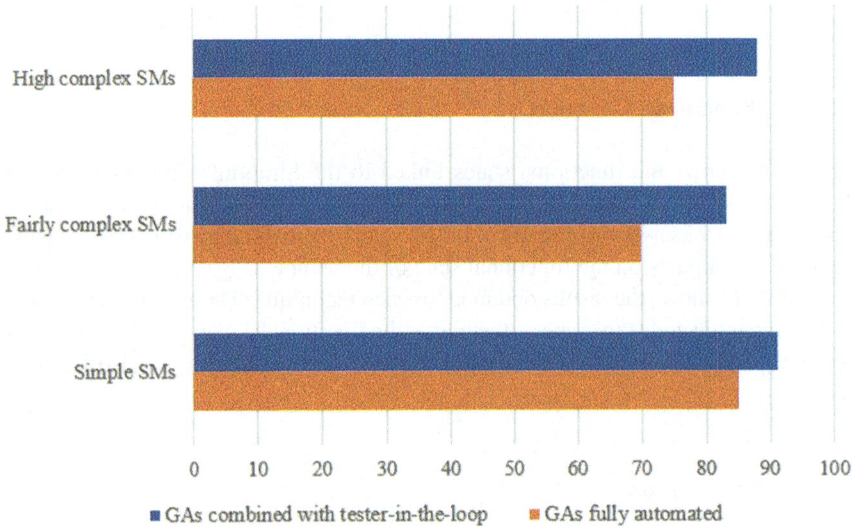

Fig. 2.17 Success rate when automating the HIL simulation

2.3.3.4 Bugs

Types of Bugs

Generally, all techniques detect the same types of bugs linked to Simulink® blocks. Some examples of Simulink® blocks where a bug was found are shown in Table 2.28 and Fig. 2.18. Some types of bugs linked to multiple calculations such as temperature or gas speed estimators can only be detected when using HIL simulations combined with dlls. Figure 2.19 depicts the obtained result for a software variable output of an

Table 2.28 Types of bugs found

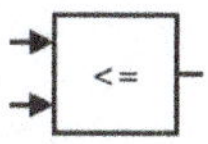	Matlab® native comparator block. It has problems in all its versions (greater than, greater than or equal to, less than, less than or equal to). In engine ECU software, on many occasions the value of a certain physical magnitude (e.g., motor revolutions, vehicle speed) is compared with a calibration threshold
	This block allows choosing between two possible paths
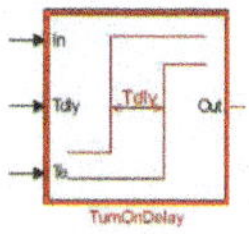	This block sets the output to TRUE while the input In remains TRUE for a certain calibratable time. Otherwise, the output is FALSE. As found in this research, when it comes to average and complex SMs, it is more difficult than in simple SMs to succeed by making the input In remain stable
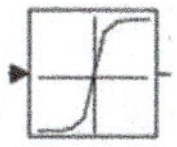	Interpolator block. In this case, depending on the input values presented in the Simulink® block, an output value is provided by applying an algorithm or an interpolation method
	The Saturation block produces an output signal that is the value of the input signal bounded to the upper and lower saturation values. The upper and lower limits are specified by the parameters Upper limit and Lower limit
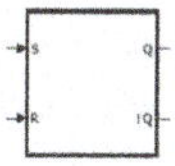	This block works as a typical RS flip-flop. As in a falling edge block, when it comes to average and complex SMs, it is difficult in certain cases (for example when validating exhaust gas treatment systems or oil adaptive maintenance functions) to reach the conditions when the S-input could be activated
	This block provides a Boolean type TRUE when a falling edge is detected. Otherwise, it remains FALSE. In this case, when it comes to average and complex SMs, it is difficult in certain cases (for example when validating exhaust gas treatment systems) to reach the conditions to generate a falling edge

SM when running the software by using an HIL simulation (in red) and its expected value (in blue). The error between the red and blue lines, represents an inaccuracy regarding the calculation of the gas speed in the exhaust pipe, which impacts the amount of urea injected to treat NO_x. Since this bug does not imply the presence of a functional bug unless it causes a malfunction detected by the driver, it is not detected by using the cause-effect technique or the model-based testing one. Only GAs combined with Simulink® model can detect it.

Number of Bugs

The results are shown in Fig. 2.20. GAs overperform the rest of the techniques used in this study because Simulink® blocks are used as shown in Fig. 2.13. Regarding the model-based testing, the fact of using models ensures better results than the cause-effect technique. Finally, the cause-effect technique performs least efficiently as no

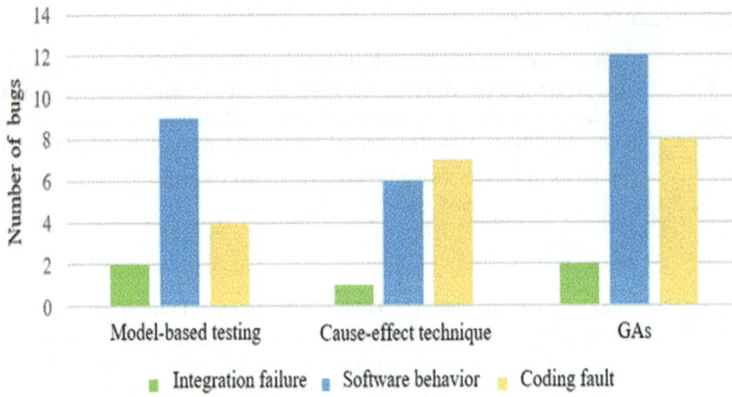

Fig. 2.18 Types of bugs found

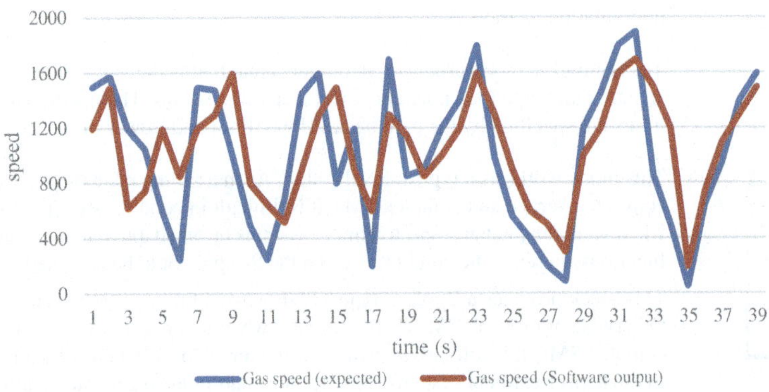

Fig. 2.19 Bug not detected unless GAs are used

model is used. The result is that it is extremely difficult to establish both the code and functional coverage.

2.3.4 Discussion

2.3.4.1 Test-Case Formulation

Several challenges must be considered when designing test-cases.

a. The engine ECU software consists of SMs composed of an important number of inputs and outputs which are usually analogical. Consequently, their values range between specific intervals. When running test-cases, it is difficult to reach

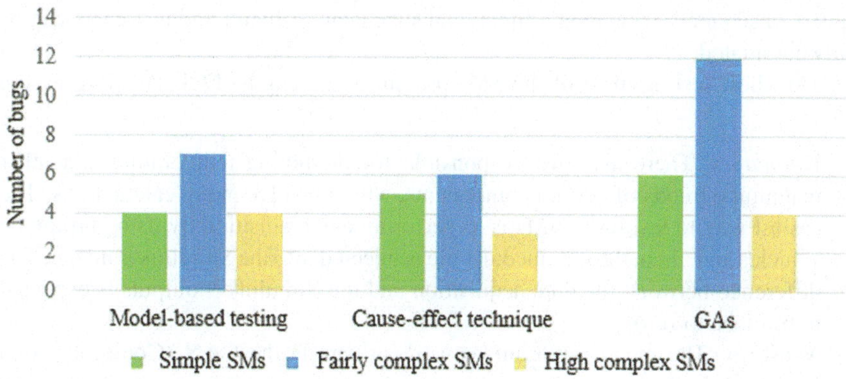

Fig. 2.20 Number of bugs found by using each technique

values contained in the variable range. For example, a variable representing the soot present in the diesel particulate filter can take a value of 40 g.

b. Considering the number of variables of SMs and their ranges, it is not possible to generate and run all test-cases which could cover the whole combination of the spectrum. In some occasions, when a variable takes a value close to its upper limit, the test-case can fail. However, if values are not close to this value, the test-case provides the expected results. That is why, at least during the validation process, the functional model must be covered with different test-cases which take different combinations.
c. Constraints must be considered to avoid generating uncoherent test-cases (for example speed = 100 km/h and first gear engaged).
d. When running a test-case by using automation processes, it is not possible to obtain the exact values indicated in the test-case due to SM interactions. Thus, the expected outputs specified in the test-case might be no longer valid. Therefore, the traditional formulation of test-cases based on input and expected output values cannot be used in simulations. Dlls allows solving this technical issue as depicted in Fig. 2.5. Thanks to them, it is always possible to assess Eq. (2.2) as they can provide the output values for the input ones reached during the HIL simulation. Therefore, GAs can check if software runs as expected by comparing the HIL results and the Simulink® models results.

2.3.4.2 Test-Cases Automation

Python scripts for automating the process must keep the inputs of the SM in a specific range. Otherwise, the expected output of the test-case may be no longer valid (Fig. 2.16). Regarding fairly complex SMs, as the number of variables present in SMs is high, it is recommended to use the tester-in-the-loop. High complex SMs have many functional states linked to the number of kms covered (example oil dilution rate). Consequently, reaching a functional state is not difficult and test-cases can be fully automated.

GAs allow testing most of the SMs present in the engine ECU software except for:

a. *Estimators*. There are SMs responsible for predicting temperature and other magnitude trends of certain components, which involve many calculations. The easiest way to test these SMs is to perform data acquisition by using prototype vehicles and, then, the obtained.dat file is injected into the Simulink® model. The difference between the data acquisition and the Simulink® outputs is expected to be close to zero.
b. *Networks*. The most important network in cars is the CAN (Controller Area Network). In these cases, the testers have to verify if frames are transmitted and received properly, how the engine ECU reacts when receiving an invalid value or an absent frame, etc. This statement can be applied to other types of networks. It is easier to validate networks by using the HIL simulations than using prototype vehicles.
c. *SMs which are not modeled by using Simulink®*. Dlls must be used if GAs are applied. Not all SMs of the engine ECU software have a specification based on Simulink® model. Consequently, GAs cannot be applied to any SMs. However, only 7% of the SMs did not have Simulink® models.

Certain high complex SMs need to cover many kilometers to reach the specific operating point indicated in the test-case. When validating the software, GAs cannot be used as the number of generated populations is not compatible with the project planning. In these cases, the cause-effect technique is recommended to reduce the validation time. Anyway, these SMs can be validated by using GAs if the calibration dataset is modified in the same way as it is done in this study.

2.3.4.3 Means Used to Validate

Using the most adequate means to validate is an essential topic as:

a. The difficulty to reach an operation point depends on the means used to validate. It is easier to use test failures on a probe by using the HIL model than by using a prototype vehicle. If the wrong means is chosen, many attempts are required to run the test-case properly.
b. The chances to find more bugs than by using other techniques are increased as the validation time is reduced. Consequently, test-engineers have time to run more test-cases than other techniques. Thus, the code and functional coverage are increased. In addition, implementing a model by using the model-based testing and GAs reduces redundancies in test-cases.
c. The productivity gain obtained thanks to GAs has an important impact on software quality. As shown in Fig. 2.21, if software version A is validated with some delay (weeks 17 and 18), after the specifications for software B are sent to the

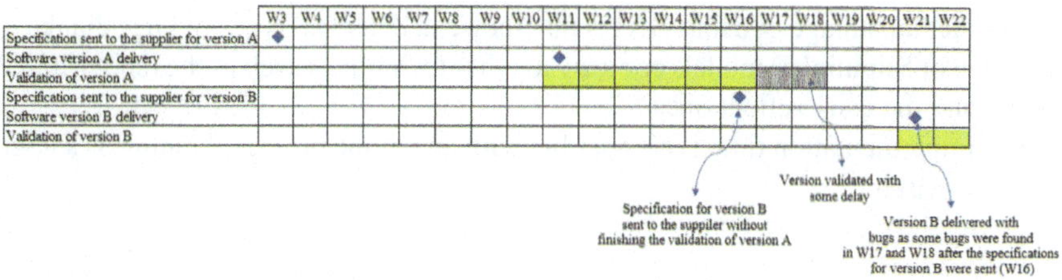

Fig. 2.21 Delays in software validation and impacts on software quality

supplier (week 16) in charge of coding the software, the software version B is delivered with bugs found in weeks 17 and 18, which may be blocking points. Therefore, the software version B could be not usable.

d. The test-engineers establish the best means according to their experience when using the model-based testing and the black-box technique. Regarding GAs, a multidisciplinary team sets the cost of the functional model.

2.3.5 Conclusions

Engine electronic control unit (ECU) software is one of the most complex software which is in charge of controlling the engine as well as other systems such as exhaust after-treatment systems. Among the main issues that test engineers can face is how to choose the best means to validate (HIL simulations or prototype vehicles) as well as design test-cases which are representative enough.

This research uses two GAs to establish the best means to validate SMs and to generate test-cases in which the expected outputs are no longer needed thanks to the usage of Simulink® models used to develop the engine ECU software with the aim of improving code and functional coverage, software bugs, test-case automation capacity and productivity. The obtained results were compared with the ones got by using traditional techniques such as the model-based testing or cause-effect ones.

The results obtained in this research show that GAs can find similar results for simple SMs and high complex ones. However, when it comes to fairly complex ones (the ones that are more present in the engine ECU software), GAs perform better than the other techniques as at least 7 more bugs were found. When it comes to functional and code coverage GAs perform better. When it comes to functional coverage, GAs improve up to 11% in fairly complex SMs and 8.4% for high complex SMs when using the cause-effect technique. When it comes to the model-based testing technique, GAs improve up to 4% in fairly complex SMs and 3% for high complex SMs. The code coverage is also improved by GAs reaching 12.8% and 7% for fairly complex and high complex SMs respectively when using the cause-effect technique. When using the model-based testing, GAs perform better up to 7.1% and 1.4% for fairly complex and high complex SMs respectively.

Another advantage of using GAs is that they can detect all types of bugs thanks to the usage of Simulink® models contrary to other techniques such as the model-based testing and the cause-effect ones.

The implementation time is compatible with an engine project planning as shown in this research.

2.4 Application of Rule-Based Expert Systems in Hardware-In-The-Loop Simulation. Case-Study: Software and Performance Validation of an Engine Control Unit[5]

2.4.1 Introduction

2.4.1.1 Background

Innovative techniques to validate software are needed to reduce cost and increase software quality.

This research aims to check if two rule-based EXs combined with dlls perform better than other techniques widely employed in the automotive sector when validating the engine control unit (ECU) software by using a HIL simulation.

To perform this research fifteen SMs of different complexity were chosen to be validated in an HIL simulation by using different techniques such as the manual execution, the tester-in-the-loop, the model-based testing, a rule-based EX and the combination of two EXs to establish the code and functional coverage, the productivity gain, the number of bugs found, potential limitations of each technique and the success rate of the HIL simulation. The test-cases used are described in-depth in the method section.

The enhancement, that dlls and EXs offer, depends on the number of states in the functional model used in the EXs and the number of subintervals in which the SM inputs can be divided. A range between 6 and 16 more bugs can be detected when using two EXs. The HIL enhancement can reach 6%, 16.8% and 18% depending on the SM complexity.

2.4.1.2 Engine ECU Software

The electronic architecture of today's vehicles is extremely complex. As a result, the number of ECUs present in vehicles is increasingly high [1, 2]. This trend will continue in the next years, thanks to driving assistance systems, which are essential for

[5] Extracted form *Journal of Software: Evolution and Process*. 2020, Volume 32, Issue 1. https://doi.org/10.1002/smr.2223, https://onlinelibrary.wiley.com/journal/20477481.

autonomous cars. ECUs are composed of hardware and software whose complexity depends on the function carried out in the network. Therefore there are multiple software running simultaneously and coexisting in a commercial car [5, 35]. This fact forces manufacturers to improve the software quality and the validation processes [3]. In addition, it is not difficult to find estimates that indicate that the total number of lines of code present in the software ECUs of a vehicle can reach up to more than 100 million. In the future, these figures will even grow significantly up to 200 or 300 million in autonomous vehicles.

Powertrain control is a system in charge of transforming the driver's will into an operating point of the powertrain according to the performance established for the product (eg, consumption and emissions) [69]. The key element of the control system is the engine ECU composed of complex hardware and software. The hardware is responsible for getting information from sensors after a filtering process to reduce noise in signals. The software processes all data received and handles actuators to reach the operating point. In addition, when a vehicle is in motion, the engine ECU (hardware and software) interacts with other ECUs to ensure the proper functioning of the car. This implies that each ECU should receive the information at a specific time. Therefore, the engine ECU (hardware and software) must be validated to assure that engine is properly controlled, the interaction with the rest of the ECUs is rightly performed, and the passengers' safety is insured. Otherwise, some failures could occur and lead to the situation in which the vehicle stalls. This fact makes the most safety critical parts of the software a hard-real-time (HRT) system. In other words, the system is subjected to real-time constraints in which every critical task must be executed at a specific deadline to ensure the correct operation of the system. Thus, one can deduce that the software validation process is complex and needs improvements with the aim of reducing costs, increasing productivity and reliability in the automotive sector.

This chapter is focused on the engine ECU software validation (one of the most complex software present in a vehicle) and shows solutions to the main difficulties associated with traditional software validation techniques. The solution proposed is showing that two EXs working in cooperation and combined with dynamic-link libraries (dlls) perform better than traditional techniques such as the model-based testing or tester-in-the-loop among others.

2.4.1.3 Techniques Currently Used

The engine ECU software validation is based on HIL simulation, combined with different techniques for generating test cases. Three key stages must be considered when performing an HIL simulation: test-case generation, test-case execution, and validation of the execution results.

One can find different definitions for the black box concept such as "*the black-box testing is a method of software testing that examines the functionality of an application without peering into its internal structures or workings*" [70, 71]. Among others, there are three types of techniques used when applying the black-box one:

a. Equivalence portioning

The inputs of the SM under validation are divided into partitions, and after having selected representative values for each partition, the test case is conducted. Then the software behavior is analyzed. The model-based testing can be defined as the automatic generation of software test procedures, using models of system requirements and behavior. To do this, a functional model must be implemented. This technique may be considered in this research as an equivalence portioning technique in the black-box testing. Because test cases are derived from functional models and not from the source code, the model-based testing is usually seen as one form of the black-box testing. The main advantage of this functional model is that all functional states and the transition from one state to another are indicated. Thanks to this, it is easier to assess the functional coverage as the number of states covered when validating an SM is known.

The EX combined with dlls consists of using an EX to assess if the software behaves as expected. The EX is built by using rules coming from the specifications and software requirements. The dlls are the Simulink model of the SM under validation that allows calculating the software outputs when performing the HIL simulation despite the SM interactions. This topic is analyzed in-depth in this research. The authors have considered this technique as an equivalence portioning one as it is exposed in this chapter.

b. Boundary value analysis

Boundary values for the SM inputs are determined and the test-case obtained is performed. Then the software behavior is analyzed.

c. Cause-effect technique

In the automotive sector, the test engineer usually has to validate cause-effect test cases that come from the software requirements. As a result, given a series of specific causes (conditions related to inputs), the validation process has to check the effect (software behavior). An example of a possible test case could be: “In case of an ESP frame is absent, the stop and start function must be inhibited.” The tester-in-the-loop, the manual execution, or automated can be considered as cause-effect techniques in this research.

All techniques that may be used to validate the engine ECU software have to face several issues such as the SM interactions that prevent reaching the values established in the test case, the type of bugs that can be found, and the problem of enhancing the code and functional coverage. Considering that the engine ECU software has up to 70 complex SMs, the interaction between SMs is continuously present and disturbs the validation process such as electronic noise. Consequently, given a test case, it is almost impossible to make the inputs reach the desired value. The main consequence is that the expected output set in the test case could be no longer available.

Some types of bugs are extremely difficult to detect by using HIL simulation unless a technician uses a significant amount of time to analyze the data acquisition. Figure 2.22 shows an example, the obtained result for an output for a variable of

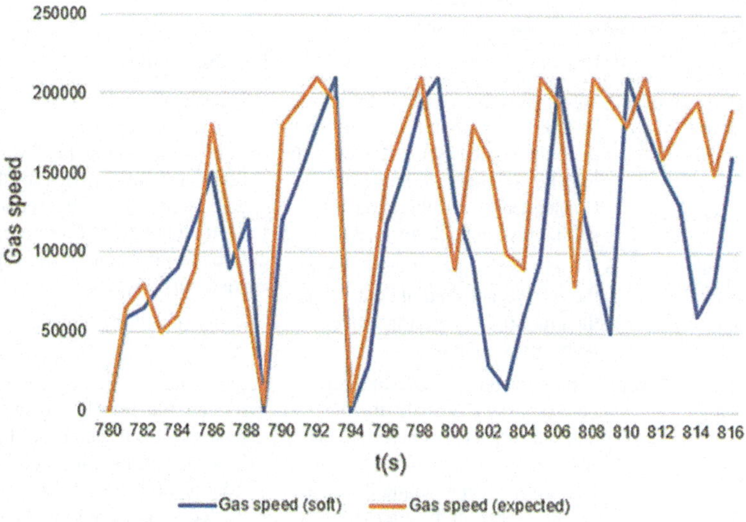

Fig. 2.22 Bug not detected when using black-box technique

an SM when running the software in an HIL simulation (in red) and its expected value (in blue). As one can see, the results are different. This error represents an inaccuracy when it comes to calculating the gas speed in the exhaust pipe. This error could impact the amount of urea injected to treat NO_x. Because this bug is not linked to a functional bug, it is impossible to detect it by using the black-box technique. The detection of this type of bug involves checking and detailed analysis of the software code by running additional software.

Considering all aforementioned, the main limitations associated with these techniques currently used in the automotive sector when using the HIL simulation are depicted in Table 2.29. The aim of this research is to solve all these limitations by using two EXs working in cooperation combined with dlls. The fact of using two EXs allows improving the code and functional coverage and gaining a better control of the automation process, thanks to dlls. It also provides an opportunity to detect any type of bugs.

2.4.1.4 Related Works

The engine ECU software validation is based on HIL simulation. Several stages must be considered when performing an HIL simulation such as test-case generation and test-case execution.

A test-case consists of a set of inputs and their expected outputs that the software should provide when working properly. In an HIL simulation, a test case is run, and the obtained result is compared with the expected one to check whether the software has operated properly for this specific test case [72–74]. There are many different

Table 2.29 Problems analyzed in this research

Limitations	Reason	Possible solution
Difficult to validate the software automatically	When the values set in the test case for the inputs are not reached, then the output values set in the test case may be no longer available. Consequently, no automatic validation can be performed	Dlls can perform this task as shown in this research as they recalculate the output values that the SM under validation should provide for the specific input values reached after the HIL simulation. Therefore, an automatic validation process can be carried out
Possible bug performance detection improvement	If input values are different from the ones established in the test case, then the software performance behavior is unknown	
Functional coverage unknown	A functional code coverage could be established by analyzing the black-box test cases before the HIL simulation However, when reaching different values for the inputs after HIL simulations, then the use cases tested are different from the ones planned	A performance rule-based EX can assess the functional coverage as exposed in this research. An EX can assess whether the SM under validation performances as expected or not, thanks to the rules used for its implementation. Thus, performance bugs could be detected. Considering the number of performance rules assessed, the functional coverage could be established
Difficult to detect bugs linked to SMs that perform many calculations	The calculations may be performed wrongly, but they do not imply that the vehicle behaves in such a way that the client could detect any abnormality	Dlls can perform this task as shown in this research as they can be used for checking whether the SM under validation calculates all SM outputs properly
Difficult to assess the code coverage accurately	There is no code model or something similar to use it for calculating the code coverage when using the black-box or similar techniques. It must be considered that there are many if-then structures in the software, which makes it extremely difficult to test all possible paths. However, the question is if the whole performance rules have been tested with a considerable number of software rules	A software and a performance rule-based EXs can assess the functional and code coverage as exposed in this research. It can be employed to establish the code and performance coverage

ways to generate a test case, such as assigning specific values to all inputs of the SMs under validation to cover a functional model, as exposed later in this research, or assessing the software performance when checking each software requirement [75–79]. The former is very difficult to implement owing to SM interactions, as it will be discussed in this paper. The aim of this method is to make the inputs reach specific values and check the outputs. The latter is widely used because the inputs of

SMs do not need to reach exact values but approximate ones to check the software performance. As a result, it is more flexible.

The black-box technique has been used for a long time in the automotive sector, as discussed by Conrad [80]. Despite its widespread use, it is true that it has some weak points, as discussed by Chunduri [81]. In their dissertation, they consider that test cases based on the engineers' experience usually imply gaps and test redundancies. Thus, they proposed a methodology to improve the black-box technique and the test-case generation. To do this, they proposed to work on three factors: enhancing function requirements specification, establishing traceability across test levels, and obtaining comprehensive function test-coverage information. In addition, it is essential to remark that the test-case execution must not be too time-consuming. Consequently, more test cases can be run, and the code/functional coverage is improved. Some research has also been focused on this topic. Zhou et al. proposed the optimized use of symbolic simulation with the aim of reducing the time required to generate a test case at the IEEE Conference [75]. As a result, given a model of a software function under validation, the time needed to cover the model will be reduced. Sopan-Barhate presented their theory about how to make the software validation process in the automotive sector more effective at the International Congress of Electronic Instrumentation and Control [76]. In their opinion, the main concerns linked to the software validation process are how to design representative test cases as well as how to prioritize the test-case execution based on priority levels, ensuring, at the same time, high code coverage rates. The solution proposed in their research is the use of orthogonal array testing.

Model-based testing is a good technique to test SMs, and it allows the assessment of the code/functional coverage in an easy manner. Raffaëlli et al. at the Embedded Real Time Software and Systems Conference, presented research focused on the usage of a functional model by running Matelo software [82, 83]. The aim of this research was to accurately assess the code coverage, as all branches of the model could be tested. The application in an HIL simulation for a more complex ECU, such as an engine ECU, was not shown. Perez et al conducted a review on the current state-of-the-art techniques used for the verification and validation of embedded systems, including software developed in the automotive sector [84]. Their main conclusion shows the need of further research concerning automatic validation, safety tests, and model validations. In short, these concepts are clearly linked to the test-case generation and improvement in automation processes. The aforementioned aspects are analyzed in-depth in this chapter.

There are many ways for automating HIL simulation in the market [85, 86]. The automation process is mainly based on black-box techniques such as those reported by Köhl et al: "*As a rule, the tests specified by the ECU departments are first performed as black box tests on the network system (know-how on software structures is not taken)*" [86]. At the 52nd Congress of the ACM/IEEE Design Automation Conference, Petrenko and Nguena-Timo reported the main problems and solutions associated with software validation in the automotive sector, on the basis of the experience of General Motor Research and Development staff, powertrain software validation team of General Motors, and the Centre of Montreal [87]. Their main conclusion was focused on the methodology known as the "tester-in-the-loop," in which the test engineer leads the system to a desired operation point, considered as a crucial one, with the aim of assuring the correct execution of the test case in such a way that the software behavior can be assessed. Once the crucial point is reached, a series of automated actions are executed to reach the goals previously established in the test case. Tatar and Mauss proposed at the ERTS Congress: Embedded Real Time Software and Systems the possibility of not using HIL simulation. Instead, by using a virtual platform, engine ECU software could be validated, thanks to the interaction with a car model [88]. As a result, many points could be tested. All the possible issues or bugs linked to the software integration on the hardware would not be detected. Koopman and Wagner exposed the main future issues when it comes to software validation in the Society of Automotive Engineers Congress. One of the most important concepts introduced in their dissertation was the "driver-out-of-the-loop" concept. Currently, the ECUs are validated by considering the driver's actions on the vehicle (accelerations, braking, etc.). If the vehicle is autonomous, these driver's actions are not relevant, and some external factors such as traffic and pedestrians must be considered to validate the software. As a result, they consider machine learning techniques as a key aspect in the future.

2.4.2 *Method*

2.4.2.1 Description

The aim of this chapter is to validate the following hypothesis:

> Two EXs working in cooperation perform better than traditional techniques when validating an engine ECU software. In addition, two EXs can overcome the difficulties depicted in Table 2.29.

To do this, a series of test cases are run by using the following techniques: the cause-effect one the model-based testing one, one EX combined with dlls, and finally, two EXs combined with dlls by using the HIL simulation. Then the following parameters are measured for each technique to validate the hypothesis: code and functional coverage, productivity, bugs found, and automation process success.

2.4.2.2 Data Used in This Research

The methodology proposed in this study has been tested in three types of functions or SMs chosen according to the number of calculations to be done as well as their complexity, number of inputs and outputs of the SM, and the accuracy required for the output results They have been considered as representative for this case study by the authors and the company subjected to this research. Considering the experience of the company that is the subject of this case study, three types of SMs or functions can be distinguished as shown in Table 2.2.

When generating test-cases, three strategies were followed in this research:

1. Generating pseudorandom values for the SM inputs under validation in such a way that all paths of the models that belong to EXs are covered. For each combination of the inputs, the performance EX must assess the expected behavior of the vehicle (represented by an HIL bench) in cooperation with a software EX that will cover a software model to assure a high code coverage. The right outputs for all inputs generated in the test case are known by using the dlls. All aforementioned statements are exposed in this section. In this chapter, as exposed later, manual test cases were also generated in order to cover the functional and software models.
2. The company under this case study has a database in which the staff document different bugs found throughout the engine project. The main advantage of this process is to guarantee easy mainstreaming between projects. All data stored in this database are handled in meetings with the supplier responsible for coding the software and designing the hardware on a weekly basis. Test engineers design test cases on the basis of different inputs such as this database, functional defects found during driving tests, specifications requirements, as well as the defects found when the engine has been marketed. The goal is to keep the test-case libraries as complete as possible over time. When the test engineer has designed the test-case library for a specific SM, a validation process is carried out. The test engineer and the designer of the SM verify whether the use cases presented in the test-case library are representative enough. For each of the test cases presented in the database, it is possible to assign values to the SM inputs with the aim of checking the software rules.
3. Pseudorandom values are generated by Matelo software with the aim of covering the whole functional model. It must be reminded that this technique is an equivalence portioning one. As test cases are generated by Matelo, the functional model is covered. Matelo assesses the functional coverage automatically. Matelo could also be used to implement a software model. However, authors have not carried it out in this chapter.

Table 2.30 shows the number of tests considered in this research according to the type of SM.

The difference between the number of test-cases for each type of SMs is because the fairly-complex SM involves a greater number of use-cases.

Table 2.30 Number of tests used in this research

Type of SM	Number of test
Simple	250
Fairly-complex	1,250
Highly-complex	100

Table 2.31 Two EXs combined with dlls

Technique	Method
Cause-effect technique	A1
Model-based testing	A2
One EX combined with dlls	A3
Two EXs combined with dlls	A3

A1: A database in which the staff trace different bugs found throughout a project. In addition, several test-cases come from the software requirements

A2: Pseudorandom values generated by Matelo® to cover a functional model

A3: Pseudorandom and manual values generated by Python scripts

Table 2.31 indicates the methods followed to generate test-cases for each technique.

It is important to analyze what A2 and A3 mean. In A2, Matelo can generate off-line (before the HIL starts) all necessary test cases with the aim of covering the functional model. In A3, Python scripts also generate test cases trying to cover the software model. The Python scripts generate pseudorandom values trying to reach software states not implemented in the model. A software state not implemented in the model involves a use case not considered by the design team, in other words, a design error. In addition, a test engineer generates manually off-line test cases by establishing the most likely combination of variables by using fuzzy values to cover the functional and software states. This process consists of avoiding illogical situations such as engaging the fifth shift when the vehicle is at 5 km/h. These inconsistencies must also be taken into account when generating automatically test cases by using Python scripts. The fact of using fuzzy variables, as exposed later, allows increasing the combination of the inputs of the SM under validation. These test cases generated manually are run by using Python scripts.

For confidentiality reasons, the list of test cases cannot be published. However, It is important to remark that fuzzy variables are used when using EXs combined with dlls by increasing the number of combinations of the inputs provided by the SM under validation.

2.4.2.3 Equipment

The following means used in this research are shown in Table 2.32.

2.4.2.4 Methodology Proposed

In this section, the key elements used in this technique are presented (EXs and dlls). Then, the process how they collaborate to run a test case is described.

Table 2.32 Equipment used in this research

Item	Description	Phase where the item is used	Cost
HIL Bench	HIL bench manufacturer dSpace®, model dSpace® Simulator Full-size (dSpace, 2016a). Versatile HIL simulator capable of emulating the dynamic vehicle behavior	Every time a test-case is run. Necessary for the HIL simulation no matter which technique is used	Depending on the characteristics of the HIL bench, the price can vary. Estimation for this case-study: €100,000 each bench
INCA version 7.1.9 provided by ETAS® (BOSCH) (ETAS, 2017)	Software used to make measurements of different software variables stored in the engine ECU memory	Every time a test-case is run. Necessary for the HIL simulation no matter what technique is used	The price depends on the number of licenses. For a big car manufacturer, an estimation of €5,000 for each license can be made
Matlab® R2013 and Microsoft Visual Studio 2015	Software necessary to create dlls	Every time a test-case is run and the user wants to avoid the SM interaction problem	The price depends on the number of licenses. This information was not provided by the company subjected to this case-study
Matelo®	Software used for validation purposes being able to generate test-cases	Necessary to generate test-cases when using the model-based testing technique	The price depends on the number of licenses. Estimations of 20 licenses are €100,000
ControlDesk® version 5.1 from dSpace. (dSpace, 2016c)	This software is needed to build the HIL model which belongs to dSpace® manufacturer. The HIL model was built by the company subjected to this case-study	No matter which technique is considered	No information about cost was provided by the company subjected to this case-study

a. Expert systems

Two EXs are distinguished:

- Software EX

  Its aim is to establish the software rules which must be applied to assure the software operation, such as a sequence of updating variables to be followed when a failure occurs. A software rule is a Simulink® path to be followed to reach a specific operation point.

- Performance EX

  The second EX is responsible for checking whether the vehicle responds as expected for a specific use-case. The first EX only verifies if the software rule is applied. The other one is abstracted from the software and only focuses on the fact of verifying the correct behavior from vehicle performance point of view. Properly coded software may exhibit wrong behavior owing to design errors as some use-cases were not considered in the specifications used for coding the software.

b. dlls

As exposed earlier, it is highly unlikely to reach the operation point set in the test case because of SM interactions. This fact implies that the automation process is not easy to be performed. Figure 2.2 depicts the process to automate a test case by using Python scripts. During the HIL simulation, the script is in charge of performing all the necessary manipulations on the driver-ECU interface model automatically. During this process, a data acquisition is performed by employing the INCA software. If these values are not reached after time out elapsed, the data acquisition is stopped and the dll is called. The dll represents the Simulink model of the SM under validation, and it allows assessing and providing the expected values of the SM for a specific state of the ECU. Thus, by using dlls, it is always possible to obtain a result after an HIL simulation. Thanks to this data acquisition and a C-file, it is possible to call the dll.

c. EXs and dlls working in collaboration

Figure 2.23 describes the process.

- **Phase 1**. The software EX establishes the test-case to be run. It must be reminded that a rule corresponds to a Simulink® path of the model of the SM under validation. This rule is communicated to performance EX with the aim of establishing the performance rule to be applied during the HIL simulation.
- **Phase 2**. The HIL simulation is performed trying to reach the operation point established in the test-case.
- **Phase 3**. A test-case is composed of a series of input values and the expected outputs. If the specific operation point is not met after a specific time elapsed, then the expected output set in the test-case may not be longer valid. The dll of the SM under validation allows assessing the right output values for the current engine ECU state. The software EX collects this information and assesses the software rule that was tested after the HIL simulation.
- **Phase 4**. The software EX sends a message to the performance EX about the software rule tested in such a way that the performance EX can update (if needed) the expected software behavior.
- **Phase 5**. Both EXs checked the HIL simulation results and decide whether the software behavior is correct and meet the specifications.

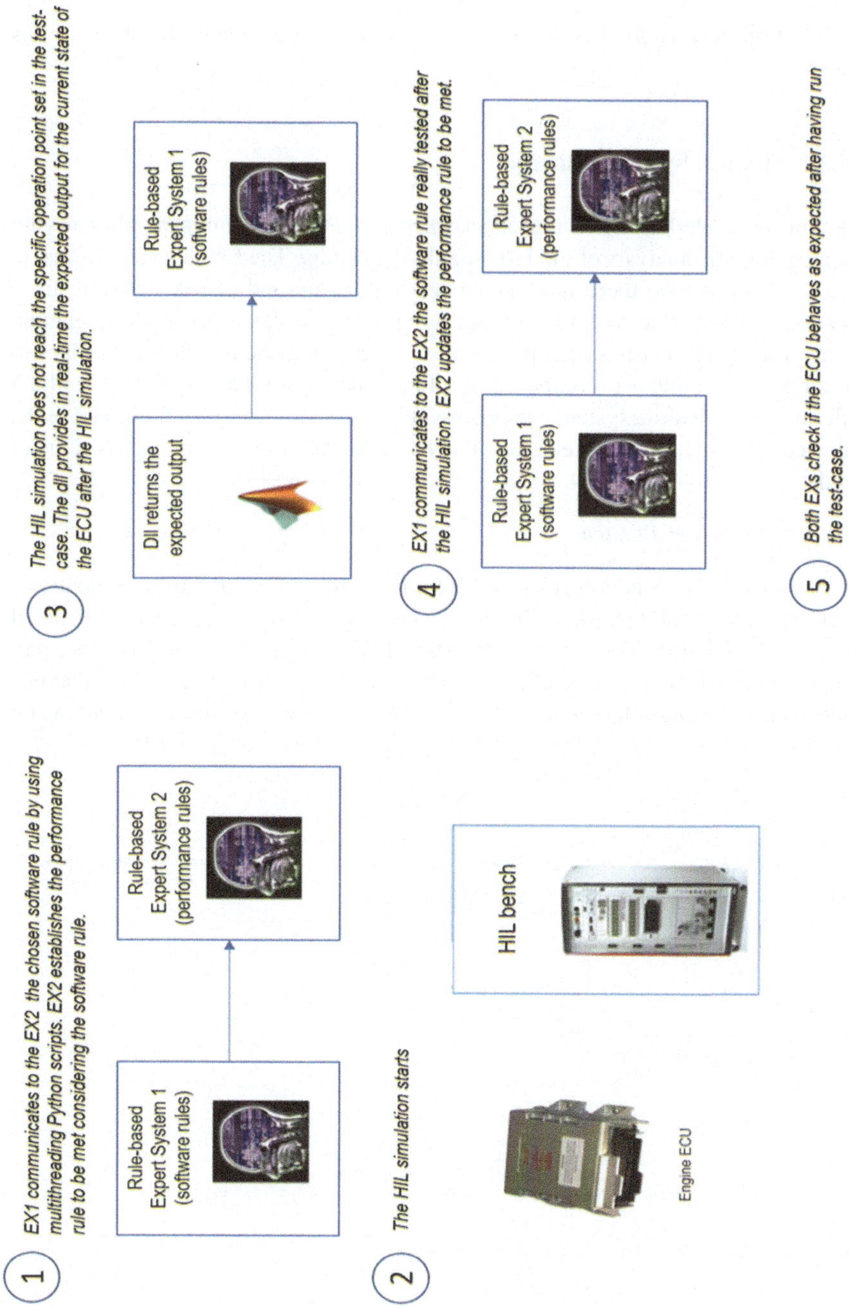

Fig. 2.23 EXs working in cooperation

2.4.3 *Validation of the Key Elements: EXs and Dlls*

This section describes the validity of the different key elements involved in this research.

2.4.3.1 Expert System Validation

The aim of the rule-based EXs is to check whether the software runs properly, carrying out an automatic analysis of the HIL simulation results. The EX design is shown in Fig. 2.24. As shown, there is a knowledge base composed of rules coming from functional or software requirements set by experts and designers at the beginning of the project. These rules are the base of the expert knowledge. When it comes to the inference engine, it is composed of a functional or software models describing different states that the system can process when applying the rules presented in the knowledge base. It must be reminded that two EXs are designed for each SM under validation.

a. Software expert system

The aim of this EX is to check whether the software meets software specifications. To better understand this, Fig. 2.25 must be analyzed. One can see a software model of a given SM, where S1–S6 represent a state. In this case, the state represents a part of the Simulink model. The conditions to be met to pass from one state to another one come from the Simulink model used to code the software. As a result, depending on

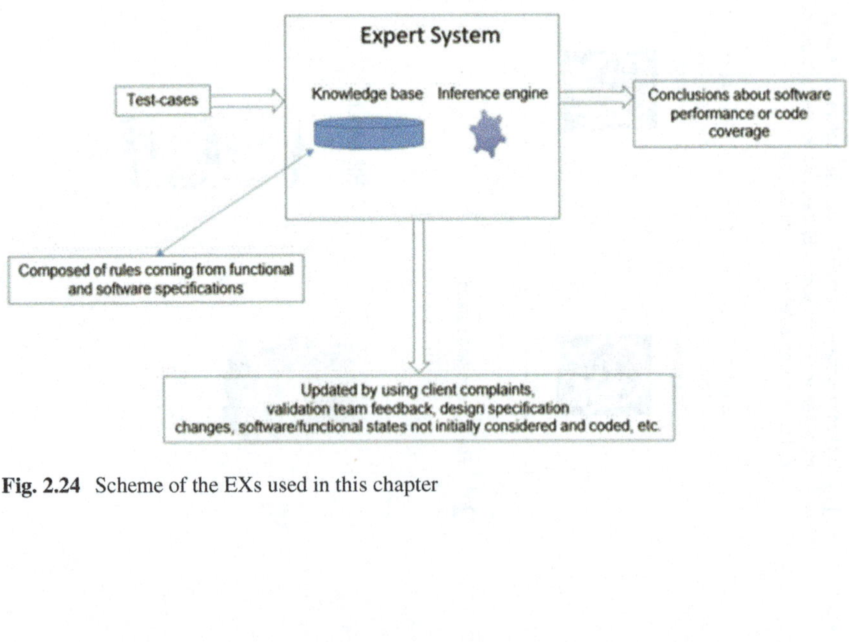

Fig. 2.24 Scheme of the EXs used in this chapter

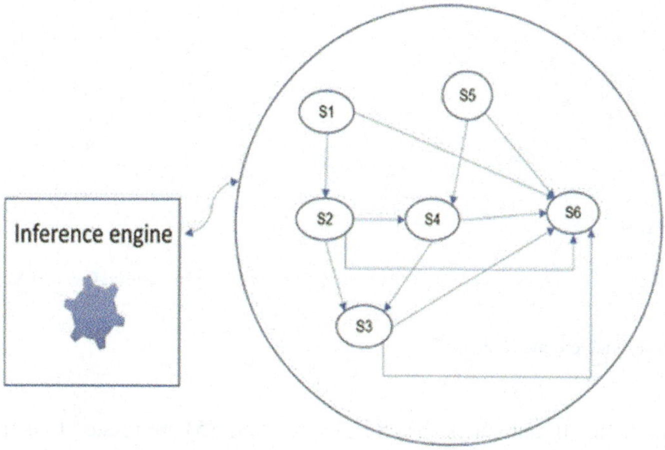

Fig. 2.25 Inference engine in detail

the HIL simulation, the values of the software variables of a given SM are analyzed in such a way that the final state is set. By checking different states covered after having executed a certain number of test cases, it is easy to have the first estimation of the code coverage. As exposed in the performance section, a test case could be run and the inference engine may not know in which software state the system is. This fact can occur, and it happens when a use case has not been considered by the design team. That is why all states in Fig. 2.25 are linked to state 6 as it represents an unknown software state.

To obtain an accurate code coverage, two key actions have been performed in this research:

- Generation of test cases in such a way that the range of possible values for a given variable is divided into intervals. In this way, the probability of covering all paths of the Simulink model is increased.
- Usage of as many states as necessary to describe the system.

b. Performance expert system

The performance EX is built by using functional states in which the vehicle can operate. Therefore, the model is not focused on part of the Simulink model of the SM under validation. The fact of covering the functional model allows assessing the functional coverage but not accurately as depicted in Fig. 2.26 when assessing the transition from S2 to S4; it is unknown if the value for Out1 was obtained following the path1 or the path2.

When a test case is analyzed by the performance EXs, after having applied different rules, the inference engine determines the state of the system. Therefore, the EXs decide whether the outputs provided by the software are coherent for the test case simulated. At this point, it is vital to verify in-depth the inference engine.

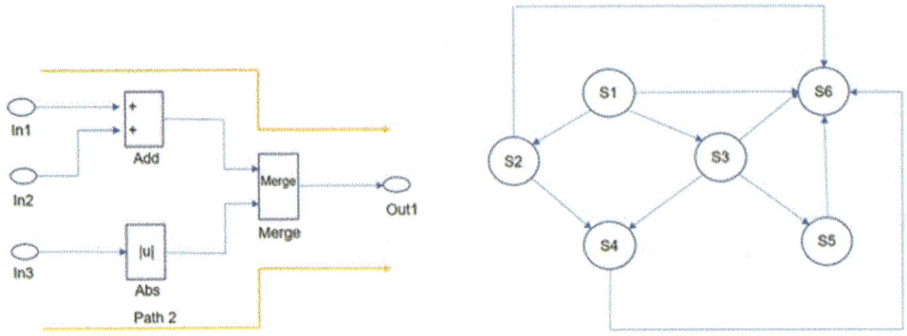

Fig. 2.26 Inference engine in detail

As shown, all functional states (S1, S2, S3, S4, and S5) are related to a state called S6. S6 corresponds to an unexpected or unknown state, which represents a use case not considered by the designers. By using this state, test engineers can improve the EXs if needed. The S6 state will be analyzed later. In this research, the EX code is not provided as it belongs to the company's know-how and is confidential.

The validation process of both types of EXs is stimulated following these two phases:

- The established rules, used by the EX, are checked following a procedure consisting of a meeting between designers and testing engineers to assure the conformity of the EX. Then, the EX is implemented by using Python.
- The aim of the validation process is to check two key characteristics: firstly, to assure that the rules presented in the knowledge base are coherent and secondly, to verify that the EXs can assess the software performance properly. To do this, a set of data acquisitions, already analyzed by test engineers, is used for the aforementioned purposes.

2.4.3.2 Dynamic-Link Library Validity

Dlls are a key element of this research. The reader may think that the fact of using dlls could keep the validation process from checking the SM interactions. This statement is not true for several reasons:

- The effects due to inputs and outputs of SM interactions are collected in the data acquisition file as it is the result of the HIL simulation.
- It is essential to distinguish some important points when it comes to designing the engine ECU software. Before integrating the software into the hardware, there is a process of building prototypes with the aim of checking whether the Simulink models work properly. Once this is checked, the decision of integrating software and hardware is made. Afterwards, the design specifications are written, all the SMs are assembled, and finally, a software is coded and the validation

process starts. Therefore, the Simulink models are the transcription of the functional specifications of the engine ECU and must be met independently of the SM interactions, hardware design, task scheduling, software-hardware integration, etc. In addition, Simulink models are tested before sending the specifications to the supplier in charge of coding the software. Therefore, for a series of given inputs, the outputs provided by the Simulink models must be equal to the ones provided by the engine ECU software when no bug is discovered. Otherwise, the functional specifications are not met.

- The fact of only considering one dll corresponding to the SMs under validation does not imply that software and hardware integration is considered as the inputs processed to the dll are the consequence of an HIL simulation. Therefore, the SM interactions are already considered in the data acquisition file. The software must provide the same output values as the Simulink model (dll). Otherwise, the functional specifications are not met.

2.4.3.3 Measurement Conditions

Before starting the HIL simulation, some conditions must be met. Otherwise, the result is rejected:

- The information provided by the probes must be equal in all cases (with and without dlls) when it comes to external factors such as air and pressure temperature and slope of the road.
- The engine ECU memory must contain no errors before starting the HIL simulation. If it does, then it must be erased by using the procedure established by the ECU supplier.
- All test-case executions must be conducted on the same HIL bench. This factor is important to assure that the same probes are being used during the whole research.

If a diagnosis defect appears when validating with dll and not when validating without dlls, or vice versa, then the test-case result is rejected and it must be executed again as the HIL model could have failed.

2.4.4 Practical Implementation

A key issue in any project is costs. Therefore, costs must be reduced as much as possible. Therefore, in this research, it has been tried to implement software validation by using Python packages. Each test case is run by using Python scripts and C-code. Firstly, the test case is performed by using Python scripts that interact with the HIL model with the aim of reaching the values established in the test case. During this process, a data acquisition is completed in ascii format. Secondly, a C-code is used to call the dlls and to assess the software behavior.

2.4.4.1 Python Scripts When Using Two EXs

From a pseudocode point of view, a multithreading implementation was conducted. One can find the thread responsible for generating software rules that will be sent to two threads: the one in charge of automation control and the one that handles performance rules (Fig. 2.27). The process is as follows. A software rule is chosen, and consistent inputs values for the variables involved in such rule are generated. Then a message is sent to the EX 2 to set the rule to be applied according the one chosen by EX 1. Once done, the automation process can be conducted using an HIL simulation. EX 2 thread is waiting for the result. The automation thread sends a message indicating if the result was correct, that is to say, whether the system reached values close to the desired operating point. If so, the EX 1 communicates to EX 2 that the selected rule was correct. Otherwise, the EX 2 updates the performance rule to be applied according to the operation point that was reached in the HIL simulation.

The second thread is in charge of controlling the automation process (Fig. 2.28), which starts when the EX 1 thread establishes the software rule to be tested (Fig. 2.28 waiting_message_from_expert_system_1). Once the process starts, the automation thread tries to lead the system to the desired state set by the EX 1 thread. The automation process ends:

- when this operating point is reached. In this case, the software and performance rules for both EX must not be updated (Fig. 2.27 automation_OK)
- when a time out elapses as the operating point is not reached because of SM interactions. In this case, the software and performance rules initially chosen might be updated (Fig. 2.26 else).

```
Thread for controling expert system 1

rule= select_rule()
data_for_automation=generate_data_for_the_rule(rule)
send_message_expert_system_2()
wake_up_thread_automation()
waiting_for_automation_result()
If automation_OK then
    send_confirmation_message_expert_system_2()
else
    send_message_expert_system_2()
end
```

Fig. 2.27 Pseudocode of software EX thread

Fig. 2.28 Pseudocode of automation thread

```
Thread for controling the automation process

waiting_message_from_expert_system_1()

while time < time_out

    control_HIL_simulation()

end

sending_current_status_to_expert_system_1()
```

Finally, thread 3 is responsible for managing the performance EX (Fig. 2.29). Its practical implementation is extremely simple, as it only runs when it is allowed by the EX 1. This can take place in two distinct situations: firstly, when the thread is instructed to select the rule to be applied according to the one set by EX 1 and secondly, when it is indicated to proceed to update the rule depending on the final engine ECU state, once the process of the HIL simulation is completed.

To implement a cross-thread communication, a submodule event from the Python threading package was chosen. Its main advantage is its ease of use. Using the wait() and set() methods, it is possible to keep a thread waiting while another performs other tasks. When the latter ends, using the set method, an event occurs to wake up all paused threads. In this case, its use is essential for several reasons:

1. The automation thread and the EX 2 threads must not start calculations until EX 1 has been initialized.
2. The thread in charge of handling EX 1 must not continue its execution as long as the automation process is finished.

Fig. 2.29 Pseudocode of performance EX thread

```
Thread for controling the expert system 2

waiting_message_from_expert_system_1()

select_rule()

waiting_message_from_expert_system_1()

update_rule()
```

3. The EX 2 thread must not continue its execution as long as a confirmation about the current status of the ECU done by EX 1 is received. The main raison is that a rule updated could be necessary.

2.4.4.2 Dynamic-Linked Libraries

The implementation of dlls allows the use of the Simulink model on multiple computers without additional cost. The dll can be implemented by following the steps indicated in many Mathworks documentation available in their site. The only thing that the user really needs is the Simulink model to be converted into a dll. In this case, these models are available as they are sent to the supplier to code the software. As described in Matlab documentation, the dll can be called by using different programming languages. In this research, C-language has been used. This process is depicted in Fig. 2.30. Firstly, when a test case is run, different software variables chosen by the user are recorded by using the INCA software. The result of this process is an ascii file that contains the variables (inputs and outputs of the SM under validation) and the specific time when each measurement was performed. Secondly, the ascii file is read by using a C-file in such a way that each line of the file is used for calling the dll (see phase 2, Fig. 2.30). The dll must return the expected output for the inputs used to call the dll. Finally, a comparison is performed as depicted in Fig. 2.30, phase 3. It must be reminded that the outputs of the SM are also available in the ascii file.

2.4.5 Results

2.4.5.1 Functional Coverage

The functional coverage would be evaluated as Eq. (2.5). This equation is widely used in the automotive sector as it allows assessing the functional coverage in an easy way by using the software requirements. Table 2.33 depicts the total number of functional requirements linked to the SMs chosen for this research.

$$FC = \frac{\text{number of software requeriments tested by a technique}}{\text{number of software requirements indicated in Table 8}} \cdot 100 \tag{2.5}$$

Table 2.34 shows the results obtained for each technique.

a. Cause-effect technique

The aim of the cause-effect technique is to check that the software requirements established at the beginning of the engine project are met. They come from a database in which the staff document different bugs found throughout the engine project. In other words, all test cases are based on the experience of the company subjected

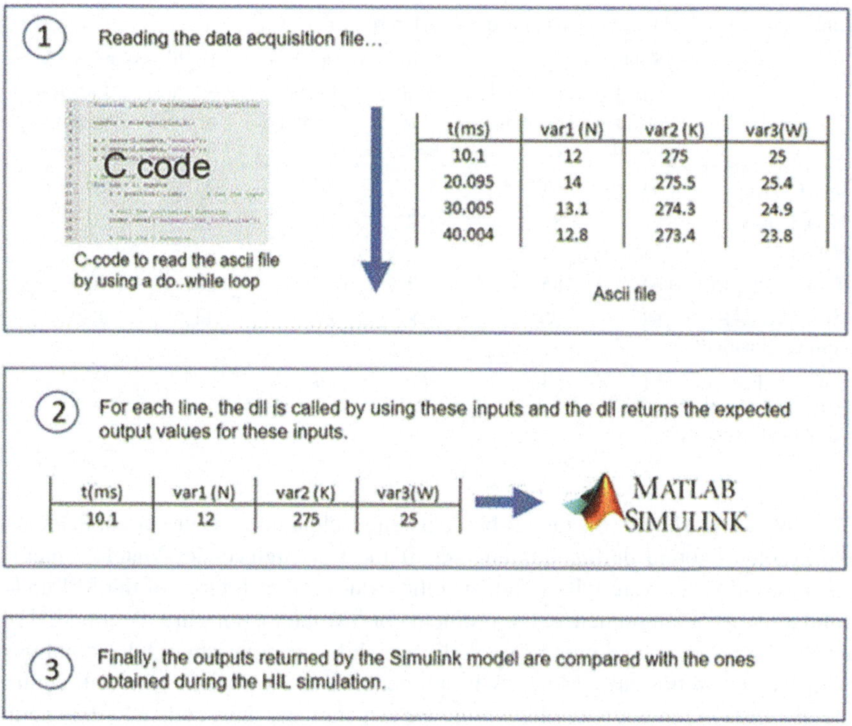

Fig. 2.30 Interactions between the C-code and the dll

Table 2.33 Number of total functional requirements

Type of SM	Number of requirements
Simple	75
Fairly-complex	400
Highly-complex	510

to this case study. These test cases can be run by using a manual execution or can be automated by employing Python scripts. The main limitation of the cause-effect technique is test-case redundancy [81]. This research confirms this statement. After having analyzed the test-cases run by using this technique, the authors found many of them which tested the same software requirements.

b. Model based-testing

As already exposed in this research, a functional model is built by employing Matelo software. In addition, this software is able to generate test cases with the aim of covering the whole functional model. The functional coverage can be calculated easily by using Eq. (2.5). Moreover, this technique allows detecting use cases not considered initially in the software requirements.

Table 2.34 Functional coverage obtained for each research

Technique	Simple SM		Fairly-complex SM		Highly-complex SM	
	Number of rules tested	Functional coverage (%)	Number of rules tested	Functional coverage (%)	Number of rules tested	Functional coverage (%)
Cause-effect	64	85.3	312	78	357	70
Model-based testing	64	85.3	312	78	357	70
Tester-in-the-loop	64	85.3	312	78	357	70
Performance EX combined with dlls	68	90.7	348	87	445	87.2
Software EX and performance EX combined with dlls	71	94.6	360	90	465	91.2

When using Matelo (the model-based testing technique), it is important to expose the problems found during this chapter. If the test engineer let Matelo generate test cases, this software will assign specific values for each input of the SM under validation. As a consequence, the problems of SM interactions are identified. That is why this strategy could not be used. To face this issue, one can use dlls combined with Matelo. In this case, Matelo will not generate the test case, but it will control the automation process. In order words, the test engineer must code a Python script to generate the test cases needed, and then Matelo will check the functional states covered as the automation is performed. In the present chapter, the test engineer codes Python scripts with the aim of running the same test cases as for the manual execution, the tester-in-the-loop, and so on.

C. EXs Combined with dlls

The software performance is assured by using an EX capable of detecting whether the software behaves properly when a test case is conducted. As discussed earlier, the unexpected behavior can come from a coding fault or design error. In both cases, the performance EX can detect them. Therefore, the results obtained when validating the EX are analyzed in this section. As done in the previous case, a validation and a test phase were performed. The main problems obtained for the former phase are depicted in Table 2.35.

When the errors indicated in Table 2.36 were corrected, the EX was assessed during the validation phase. In this case, the same number of test cases used when validating the software EX was performed. The acceptance process was the same as reported in the software EX validation process (Table 2.36).

When using a performance EX, a certain number of test cases were conducted by assigning pseudorandom values to the inputs of the SMs: 25 for 20 simple SMs, 5 for fairly complex SMs, and 2 for highly complex SMs. Table 2.37 depicts the results obtained.

Table 2.35 Errors detected when validating the EXs

Type of error	Cases	Percentage	Explanation
Wrong syntaxes	6	5.5	Because the rules used to design the EXs are extremely complex, the programmer made coding errors
Incoherence among rules	2	1.8	In some cases of wrong performance of the EX, incoherence between rules was found
Misunderstanding of technical specifications	3	2.7	Because of innovative evolutions in some parts of the engine, some technical specifications were not understood properly
Rules not coded or forgotten	1	0.9	This type of error was made owing to the same misunderstanding of technical specifications

Table 2.36 Most important points checked during the validation meeting

Most important factors considered to validate the expert system
All safety concepts (ISO 26262) were modeled or considered in the EX
All diagnoses that may be detected by the engine ECU during the validation process were considered in the EX
The number of states is considered sufficient and representative enough by the project team
All use cases are modeled and considered in the EX (a priori)
The transitions among all the states considered in the EX are defined and modeled properly
The feedback of other projects was considered in the EX

Table 2.37 Code coverage when an EX is used

Type of SM	Number of rules	Number of functional states tested not checked when using an EX	Functional coverage (%)
Simple	75	7	90
Fairly complex	400	52	87
Highly complex	510	65	87.2

When both EXs are used together when performing an HIL simulation, the final results are enhanced, as more rules are checked as shown in this section (Table 2.38). The main reason behind this fact is that the higher the code coverage of the software EX, the higher the functional coverage obtained when carrying out HIL simulations. Therefore, it is essential that they work in cooperation. Another aspect that must be analyzed is why 100% functional coverage is reached when the software code coverage is not 100%. This fact can be easily explained as a specific variable can be activated by different software paths of the Simulink model. Figure 2.31 shows how output Out1 can be activated by two different paths. That is why the functional

Table 2.38 Number of rules or functional states not checked when an EX is not used

Type of SM	Number of rules	Number of functional states tested not checked when using both EXs	Software code coverage (%)	Functional coverage (%)
Simple	75	4	94.6	100
Fairly complex	400	40	90	95
Highly complex	510	45	91.2	94

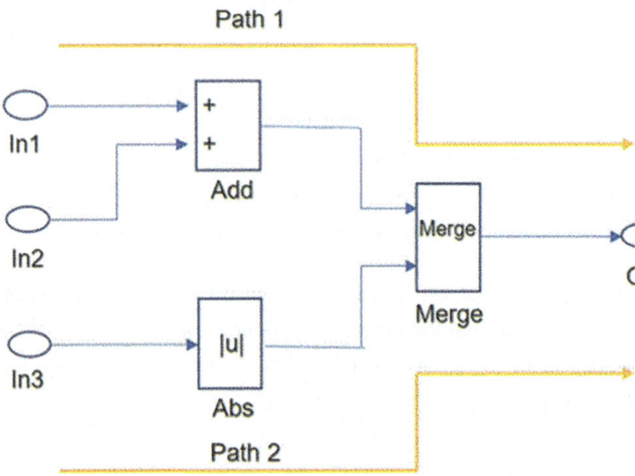

Fig. 2.31 Activation of a specific variable

coverage is 100% but not code coverage. This fact supports the conclusion that the number of subintervals is essential to get a high code coverage.

2.4.5.2 Code Coverage

The supplier responsible for coding the engine ECU software starts from the specifications composed of complex models which are provided by the car manufacturer. Thus, it is extremely difficult to reach a code coverage close to 100% as reported in previous research [81]. In order to assess the code coverage, the Eq. (2.6) was used which establishes the relation between the total number of Simulink® blocks to be tested (Table 2.39) and the total number of Simulink® blocks tested.

$$FC = \frac{\text{number of Simulink® blocks tested by a technique}}{\text{number of Simulink® blocks indicated in Table 12}} \cdot 100 \qquad (2.6)$$

Table 2.39 Number of total Simulink® blocks[a]

Type of SM	Number of requirements
Simple	75
Fairly-complex	400
Highly-complex	510

[a]When a state flow is present, each state is considered as a Simulink® block

The results obtained for each technique are shown in Table 2.40.

a. The cause-effect technique

When using the cause-effect technique, after having run all test-cases to assess the functional coverage the number of Simulink® blocks covered were calculated following the equation [2]. The cause-effect technique implies redundancies. Consequently, the code coverage is not high. The main limitation associated with this technique is that it is based on the software behavior and not on checking the code coverage and the number of Simulink® blocks covered.

b. The model-based testing

The model used for testing the SM under validation can be built from two points of view. The first one focuses on the functional software behavior. The other one focuses on the software structure, in other words, on the Simulink blocks without analyzing the purpose of each block. In this section, the second point of view is used. However, it faces the same problems already described when automating test cases because of the SM interactions.

c. EXs combined with dlls

Table 2.40 Code coverage obtained for each research

Technique	Simple SM		Fairly-complex SM		Highly-complex SM	
	Number of rules tested	Functional coverage (%)	Number of rules tested	Functional coverage (%)	Number of rules tested	Functional coverage (%)
Cause-effect	63	78.7	265	75.6	410	77.3
Tester-in-the-loop	63	78.7	265	75.6	410	77.3
Model-based testing	63	78.7	265	75.6	410	77.3
Performance EX combined with dlls	74	92.5	295	84.3	435	82
Software EX and performance EX combined with dlls	76	95	313	89.6	425	80.2

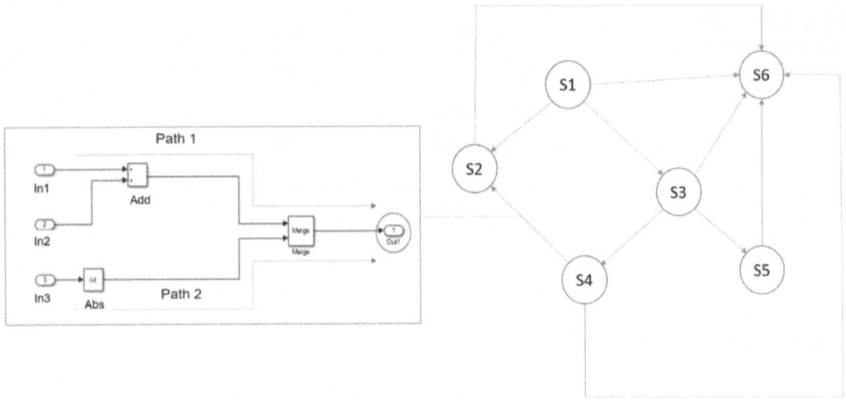

Fig. 2.32 Scheme of a software EX used in this research

A realistic way to assess the code coverage is to check whether all sub-blocks which composed a Simulink® model of a specific SM under validation, are verified after having run all the test-cases. In this research, two options were considered:

a. Division of the range of every software variable involved in the validation process into subintervals. The aim of this was to generate test-cases that allow covering as many paths of the Simulink® model as possible. This strategy is followed by commercial software such as Matelo®.
b. Number of states. This is a key factor as it allows modelling in detail the software behavior by using functional states. As depicted in Fig. 2.32, every path of a Simulink® model may be represented by a functional state.

By changing the value of these factors, the code coverage was assessed. To do this, it was checked how many functional states were covered when conducting all test-cases available to validate an SM following the strategies described earlier to generate test-cases. The obtained results are shown in Table 2.41. These figures show how the code coverage increases as the number of states goes up. This fact must be coherent with the functional coverage rate. This point will be analyzed in this section.

The code coverage could be calculated in a more accurate way. However, this implies that two main issues should be taken into account. Firstly, the number of test cases to be performed by using an HIL simulation increases, and the project time frame can be affected. In addition, some use cases are difficult to be simulated when using an HIL bench owing to the HIL model limitations, especially when it comes to SMs linked to advanced driver assistance systems. It must be reminded that these functions need a lot of information exchanged between different ECUs present in the CAN network. Secondly, the number of states should also be increased. However, it cannot be stated that the more states are used, the higher the code coverage is. As shown in Table 2.41, there is a limit at which the code coverage does not increase meaningfully (15 states for a simple function and 75 for a fairly and highly complex function). After analyzing the results, the conclusion was that many test cases were

Table 2.41 Code coverage trend depending on the number of the states (measured in %) (Subinterval = 3)

Number of states\type of SM	1	3	5	8	11	15	18	20	25	31	36	42	48	54	60	68	75	80
Simple	1.5	6.3	14	45	75	95	95.2	95.3	95.3	95.4	95.4	95.4	95.5	95.5	95.6	95.6	95.7	95.7
Fairly-complex	1.19	2.1	2.6	3.5	5.2	8.5	15	17	35.6	36.9	42.3	50	57.1	64.3	71.4	78.6	89.3	89.6
Highly-complex	1.1	1.8	2.2	3.1	4.7	6.7	13.5	15.8	29.8	33.2	38.2	43.2	53.2	58.5	68.7	72.5	80	80.2

redundant. As mentioned above, some states are difficult to reach when using an HIL simulation owing to HIL model limitations.

When it comes to subintervals breakdown, the obtained results are shown in Table 2.42. The main conclusion is the higher the number of subintervals, then the lower code coverage is, as redundancy in test cases occurs. In this research, the authors proceeded to use a fuzzy logic to establish the optimal number of subintervals. More specifically, the speed was considered as low, average, and high, the water cooling temperature low, average, or high, and so on.

Figures 2.33 and 2.34 depict the results in a more visual way.

Finally, it is essential to check the validity of the software EX. Two phases were considered: a validation and a test one. On the one hand, the former consists of verifying test-cases to assess the EX performance depending on the type of SMs under validation (60 for simple SMs, 40 for fairly-complex SMs and 10 for highly complex SMs). On the other hand, the later seeks its acceptance after having tested 30 for simple SMs, 20 for fairly-complex SMs and 5 for highly complex SMs. It is vital to remark that all the points, tested to validate the system, covered all the functional rules. Thus, the functional coverage rate was 100%. In the first phase, a 17.3% error was obtained. In the second one, 0%. As a result, the EX was validated. Table 2.43 shows the results obtained during the first phase.

Before using the EX, an acceptance process is performed, consisting mainly of a series of meetings in which some key factors are assessed. Table 2.44 depicts the most important ones. All the factors assessed cannot be indicated for confidentiality reasons. It is essential to remark that no bug or unexpected behavior of the EX was detected after its validation.

2.4.5.3 Bug Detection

When using one EX, the results obtained after executing the number of test-cases specified in Table 2.2 are shown in Fig. 2.27.

- The Cause-effect technique (automated or not) and the model-based testing one.

  The use of Python scripts is a less efficient technique because it is complicated to make the system reach a specific operating point, especially when dealing with certain SMs, such as those related to after treatment of exhaust gas systems. It must be reminded that these SMs perform multiple complex and accurate calculations. As a result, this technique faces the SM interaction problem. Despite this, a test-case can be executed by using an HIL simulation thanks to dlls. This statement is also true for model-based testing. The fact of reaching specific points remains difficult due to the SM interaction problem.
- The tester-in-the-loop technique and the manual execution one.

  The tester-in-the-loop technique offers better results as a technician or a test engineer can make the system reach a specific operating point. Then, a script is run to use all the necessary manipulations on the HIL model to end the test-case

Table 2.42 Code coverage trend depending on the subintervals and the number of states (measured in %)

Number of states\type of SM	1	3	5	8	11	15	18	20	25	31	36	42	48	54	60	68	75	80
Subinterval = 4																		
Simple	1	4	8	35	68	85	85.2	85.3	85.3	85.4	85.5	85.5	85.6	85.6	85.6	85.7	85.7	85.7
Fairly-complex	1.2	2.1	2.3	3	4.2	7	12	14	30.2	31.2	36	45	52.2	56.3	62.3	70	83.2	83.8
Highly-complex	1.1	1.2	2	2.5	4.1	5	11.6	12.5	18.2	23.5	29.2	34.5	40.2	43.5	48.5	53.1	58.2	58.8
Subinterval = 5																		
Simple	1	3.5	7.2	30	51	72	72.1	72.2	72.5	72.6	72.6	72.6	72.7	72.8	72.8	72.8	72.8	72.9
Fairly-complex	1.18	1.9	2.2	2.8	3.5	6.2	10.8	12.5	25.6	28.6	32	40	48.5	53.5	58	64.5	72.5	72.9
Highly-complex	1.1	1.8	1.9	2.2	3.5	4.2	7	8.5	12.7	16.5	20.1	24.2	27.6	32.5	37.5	42.8	50	50.6

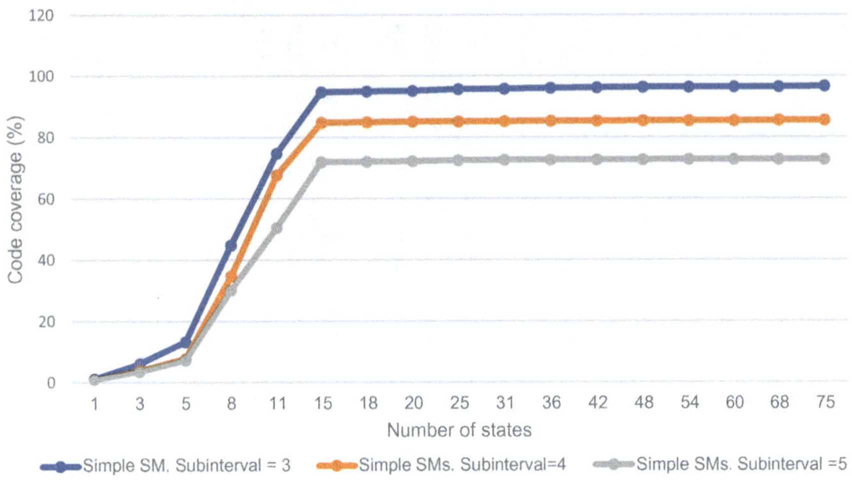

Fig. 2.33 Code coverage rate versus the number of subintervals considered when validating a simple SM

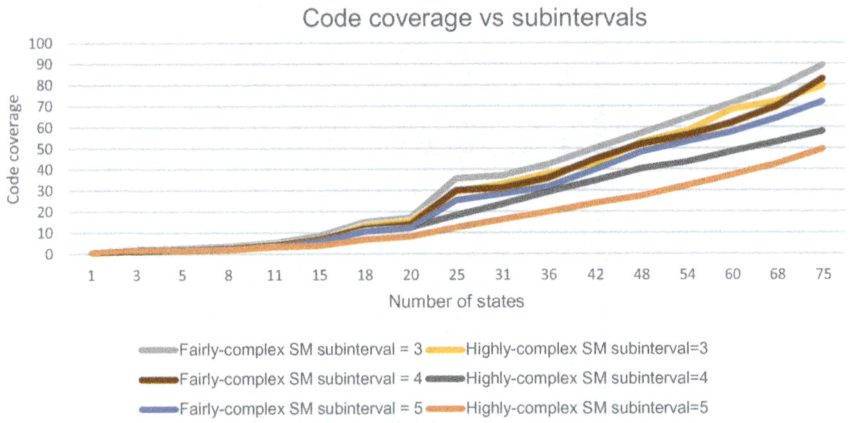

Fig. 2.34 Code coverage trend vs the number of sub-intervals chosen when validating a fairly and highly complex SMs

performance. This statement is also true for manual execution as a technician performs the whole test-case execution.

- Using EXs to validate the software

  EXs performance must be analyzed. In the previous research, which is under consideration for publication, the authors probed how the use of a performance EX introduced significant advantages such as the capacity of detecting more bugs than other techniques. The question that might arise is if the addition of a software EX introduces significant improvements, which would justify its implementation.

Table 2.43 Errors detected when validating the EXs

Type of error	Cases	Percentage	Explanation
Wrong syntaxes	10	9.1	Because the rules used to design the EXs are extremely complex, the programmer made coding errors
Incoherence between rules	6	5.5	In some cases of wrong performance of the EX, incoherence between rules was found
Rules not coded or forgotten	3	2.7	This error is due to the same misunderstanding of technical specifications

Table 2.44 Most problematic Simulink blocks

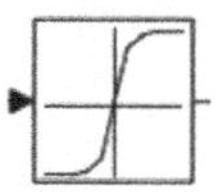	Interpolator block. In this case, depending on the input values presented to the Simulink block, an output value is provided by applying an algorithm or an interpolation method
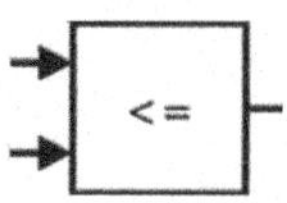	Matlab native comparator block. It has problems in all its versions (greater than, greater than or equal to, less than, less than or equal to). In engine ECU software, on many occasions, the value of a certain physical magnitude (eg, motor revolutions and vehicle speed) is compared with a calibration threshold

As shown in Fig. 2.35, the answer is yes, as more six bugs were found. This fact supports the results shown in Table 2.38; the higher the code coverage, the more functional states are checked. Six bugs were detected by using two EXs. Figure 2.36 depicted a classification of these bugs. The term of strategy chosen showed in Fig. 2.15 refers to the ability of testing more paths of the Simulink models, thanks to the use of software EXs that allow to increase the code coverage rate. The rules not considered concept refers to functional states reached during HIL simulations that had not been considered by the design team. The value bugs term refers to certain bugs detected when a Simulink block did not perform some calculations properly (Table 2.44).

2.4.6 Dynamic-Link Libraries

The problem of SM interactions is resolved, thanks to the usage of dlls as proved in this research. It must be remarked that the obtained results are very similar no matter what technique is used provided that dlls are implemented as depicted in Tables 2.45, 2.46, and 2.47.

Several factors must be considered to better understand these results. Firstly, dlls are not needed when using the manual execution as the test engineer can control accurately the automation process. Secondly, the results for "Automated with a Python

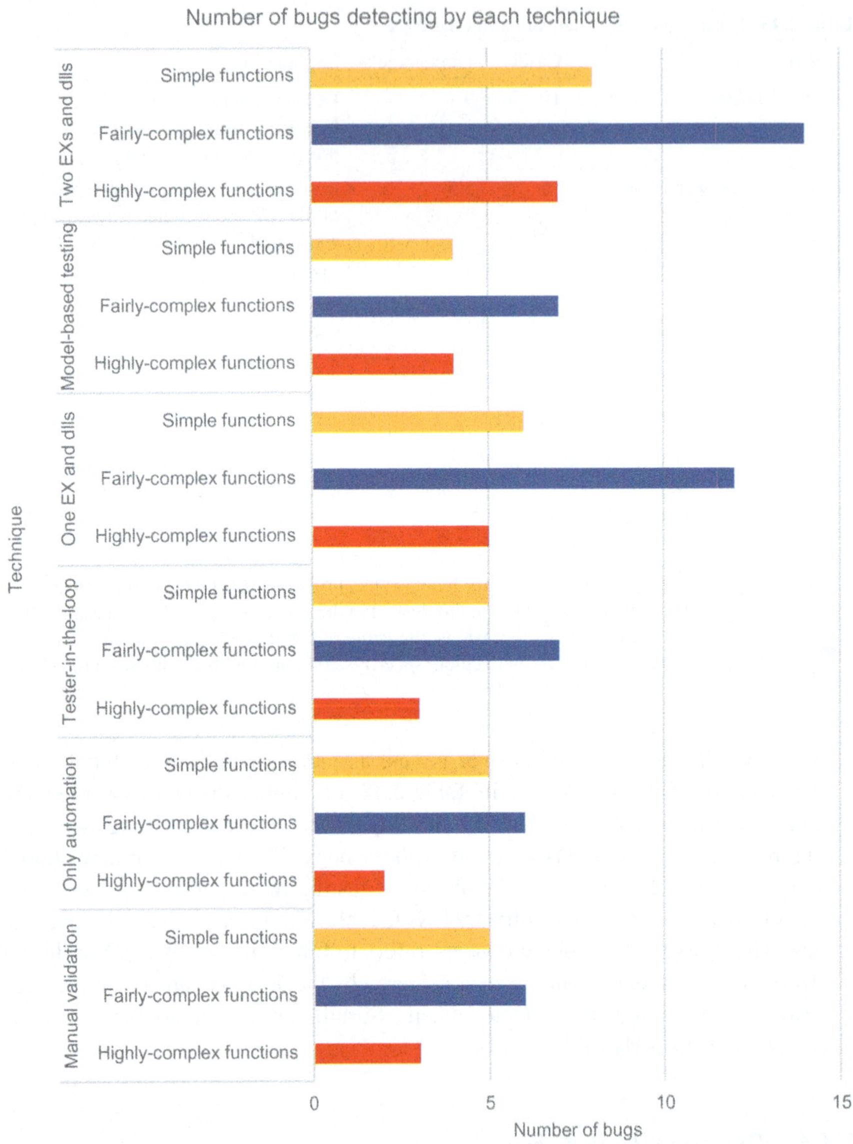

Fig. 2.35 Capacity of bug detection

script and the use of dlls" are representative for no matter what technique is used, which implies that a Python script is run to perform the automation process such as the model- based testing and EXs. Finally, when using dlls, a 100% success rate is not achieved because of HIL model inaccuracies. The HIL model, which represents the vehicle dynamic, is not perfect. Therefore, from time to time, the engine ECU

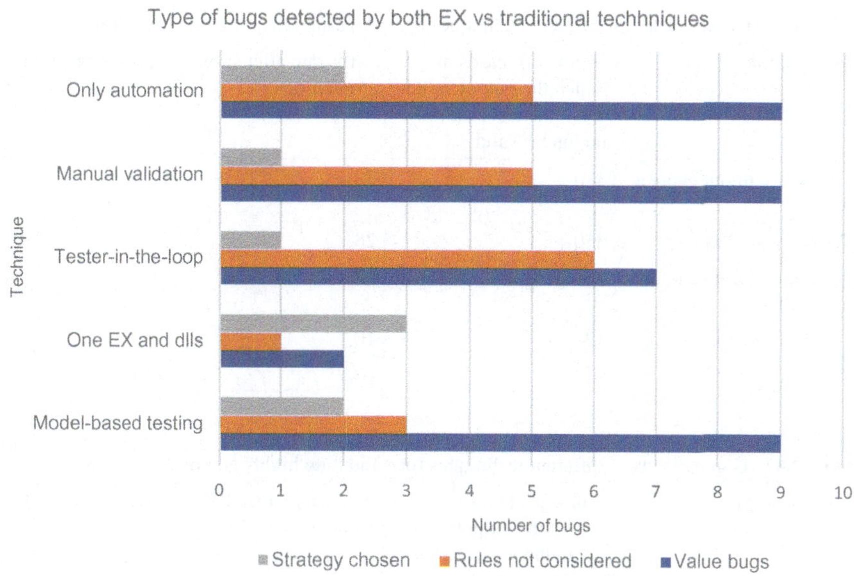

Fig. 2.36 Types of bugs found

Table 2.45 Comparisons of different techniques for validating simple functions

Methodology	Number of cases in which the output value set in the test case was no longer valid	Error rate after 250 simulations (%)	Success rate (%)
Only but without using a dlls	49	19.6	80.4
Tester-in-the-loop	5	10	90
Only automation and the use of dlls	13	5.2	94.8
One EX and dlls	12	4.8	95.2
Two EXs and dlls	13	5.2	94.8

can detect failures, which implies that the test case cannot be properly run despite the dlls usage.

2.4.7 Limitations

It is important to emphasize that the use of EXs does not allow the detection of any type of bugs. Indeed, the output provided by the software for a particular variable

Table 2.46 Comparisons of different techniques for validating fairly complex functions

Methodology	Number of cases in which the output value set in the test case was no longer valid	Error rate after 1250 simulations (%)	Success rate (%)
Only but without using a dlls	480	38.4	61.6
Tester-in-the-loop	350	28	72
Only automation and the use of dlls	125	10	90
One EX and dlls	126	10.1	89.9
Two EXs and dlls	124	9.9	90.1

Table 2.47 Comparisons of different techniques for validating highly complex functions

Methodology	Number of cases in which the output value set in the test case was no longer valid	Error rate after 100 simulations (%)	Success rate (%)
Only but without using a dlls	61	61	39
Tester-in-the-loop	35	35	65
Only automation and the use of dlls	15	15	85
One EX and dlls	15	15	85
Two EXs and dlls	14	14	86

differs from the one expected. However, if this fault does not introduce any serious malfunction, the EXs will not be able to detect it. That is why, the use of the dlls is essential in this methodology. This type of bugs may be present in SMs that perform many calculations.

The reader might think that, in case of bugs in the Simulink model, the software will also contain these errors. As a result, no bug will be detected by using the method proposed in this research. This study has proven that this statement is true and that is why the performance EX must be used. In the engine ECU software, when some specific failures are detected, a software reset takes place. If, despite this, the failure still occurs, the ECU stops the car. Figure 2.37 shows a bug found during this research. The dll and the software did not increase a counter properly. The main consequence was that instead of counting until four software resets, they counted until two and the engine was not stopped. In this case, the dll and the software provided the same outputs. However, the EX detected this software bug.

Finally, the limitation associated with this methodology is no different to others that can be proposed as increasing the number of test cases to be conducted to ensure a code coverage of 100% is not compatible with the planning of an engine design project.

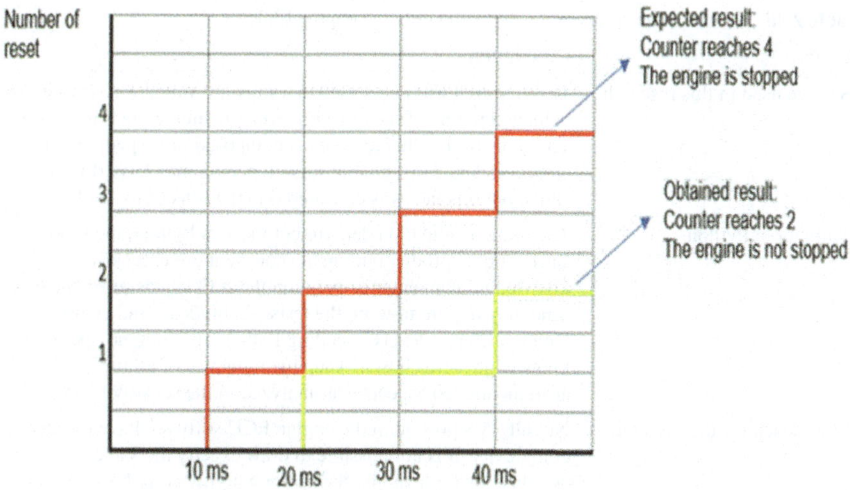

Fig. 2.37 An example of a software bug detected by the EX that could not be detected by using traditional techniques

2.4.8 *Threats to Validity*

Table 2.48 describes the main variables to be controlled (predictors) to check the influence on the response variables (productivity gain, documentation quality, and bugs). Among these predictors, one can distinguish the sample used in the study described in this chapter, the staff's skills in Python, the SM chosen to be validated, the staff's experience in how the engine ECU operates, the reliability of measures done during the validation, and finally, the quality of the documentation furnished to technician or engineers to validate the software (test description, python scripts, etc.).

All these factors are analyzed in the sensitivity analysis. The authors described in-depth how all these factors impact the time needed to code Python scripts and, therefore, productivity (internal threats). Considering that one of the most important factors to be analyzed in this research is the number of bugs found when using two EXs working in collaboration, it is essential to check how these variables impact this factor. Figure 2.17 shows that the less quality the documents have, the fewer bugs are detected, and therefore, the performance decreases. The quality depends on the sample used in this research, the training in Python, the staff's experience in the engine control unit ECU, and the number of people belonging to the staff. When it comes to external threats, it is of paramount importance to verify if the results can be generalized or if it is applicable to a larger group. Figure 2.38 shows that it can be applied as the quality depends on the number of members of the staff. This statement is based on the fact that the higher the staff is, the more hours can be devoted to improving the quality of documentation. Otherwise, the terms of the project will be prolonged.

Table 2.48 Factors to be controlled when validating the engine ECU

Factor	Description
Sample used in this research	This research was performed in a software validation service that belonged to one of the most important manufacturers in Europe. The staff used in this research is composed of 40 people: 19 engineers and 21 technicians. Each person may have different skills, but this fact was considered in the sensitivity analysis
Training in Python	The more a validation department masters Python, the more sufficient the productivity gain is or the more extensive knowledge of an engine operation the staff can acquire, the less time they require to write the tests. Technicians and engineers having different levels in coding Python or in engine operation knowledge were chosen. Then the influence of all the aforementioned aspects was analyzed in the sensitivity analysis
SM chosen for the research	Not all SMs present in the engine ECU software have the same complexity. It is not possible to draw exactly the same conclusions for a simple SM as for a highly complex one. The SMs were divided into three groups. The fact of not doing this implies that the productivity gain is not properly assessed
Unreliability of measures	All measures were taken in the same conditions. To assure this, a procedure was written, which describes when measures can be accepted and when they must be rejected it. In addition, EXs must be validated Otherwise, the conclusions could be completely random and wrong
Staff's experience in the engine ECU field	The members of the staff of a validation service may change their positions in the company. As a result, the department may have more specialized people at a specific moment and vice versa in other occasions. This research was performed considering different scenarios depending on the staff's training as shown in the sensitivity analysis
Quality of documentations provided to the technician to validate the software	A validation department can have more or less staff. It must also be reminded that a validation department is of high cost for companies, so they try to limit the number of people who run the service

2.4.9 Sensitivity Analysis

When automating a test case, it is necessary to make the vehicle reach specific operating conditions. To do this, there are two options: firstly, coding a high-quality script that can control all necessary parameters that could prevent the vehicle from achieving the desired operating point and secondly, the "tester-in-the-loop" concept can be applied. Thus, a technician makes the vehicle reach a desired operating point, and then an automation script performs all subsequent actions to run the test case completely. In this chapter, these SMs were automated in the company subjected to this case study by using Python scripts. The key to achieve this is to code libraries that can carry out specific interaction with the vehicle model interface, such as heating

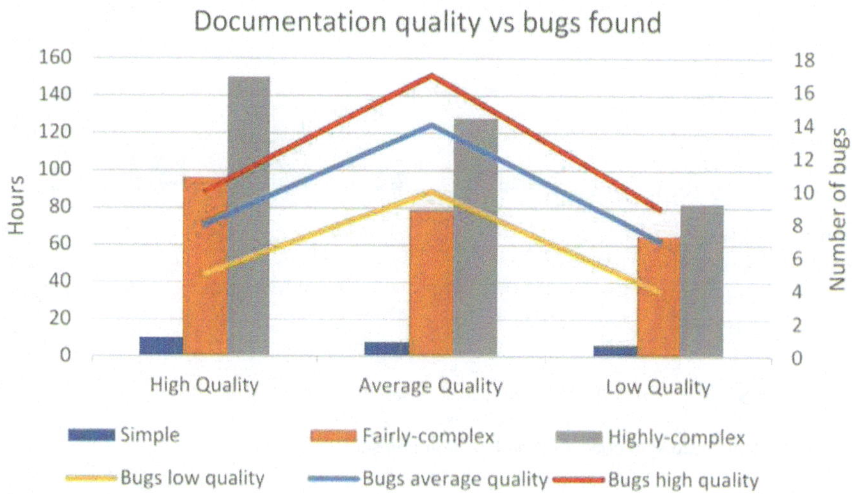

Fig. 2.38 Documentation quality vs bugs found when using Exs

the NO_x probes. Therefore, quick and robust scripts can be coded. However, the time needed to code Python scripts depends on the programmer's experience. As shown in Table 2.49, the staff of the validation software validation service of the company subjected to this case study has been classified as expert, average, and low level when it comes to their experience in Python.

Figure 2.39 depicts the obtained results.

Clearly, training in Python scripts is a key aspect to be taken into account to improve productivity when it comes to software validation.

However, training in Python is not the only key factor to improve the quality of software and time frame of the project. Knowledge about physical phenomena controlled by the SM under validation has a great influence on the time needed to design tests. For example, if a test engineer needs to design tests for validating the urea injection for the nitrogen oxide treatment, if he knows the physical foundation of the function, besides knowing the software architecture, the time needed to design a test case is reduced. To verify this, expert python test engineers were chosen to code python scripts to automate simple, average, and complex functions. However, these engineers had high, average, and low knowledge about the function to be automated. The obtained results are shown in Fig. 2.40. Consequently, in addition

Table 2.49 Staff's training in Python

Group	Experience in coding Python scripts	Number of members
Expert level	More than 2 y	10
Average level	Between 1 and 2 y	15
Low level	Less than 1 y	15

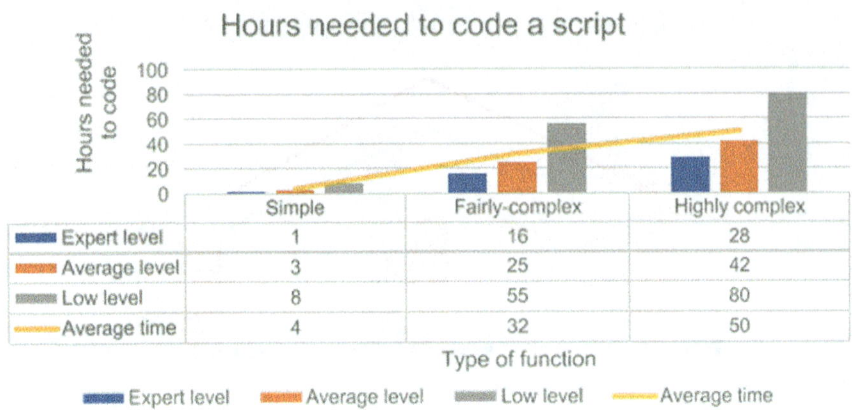

Fig. 2.39 Time needed to code a script depending on staff's training

to SM knowledge, another potential method of improvement is provided by the expertise in the physical phenomena linked to a combustion engine.

The number of engineering hours dedicated to design the tests used during the validation process depends on the final quality of the test documentation provided by the technician. If schedules, notes, and comments are attached, the cost increases. Figure 2.41 shows the total amount of engineering hours spent to design the tests depending on the final quality provided. In this research, the quality was measured by using a checklist built by the validation expert engineer of the powertrain software validation service.

Taking together Figs. 2.40 and 2.41, the total number of hours needed to design the test cases (test-case design and the time needed to code the Python scripts) is

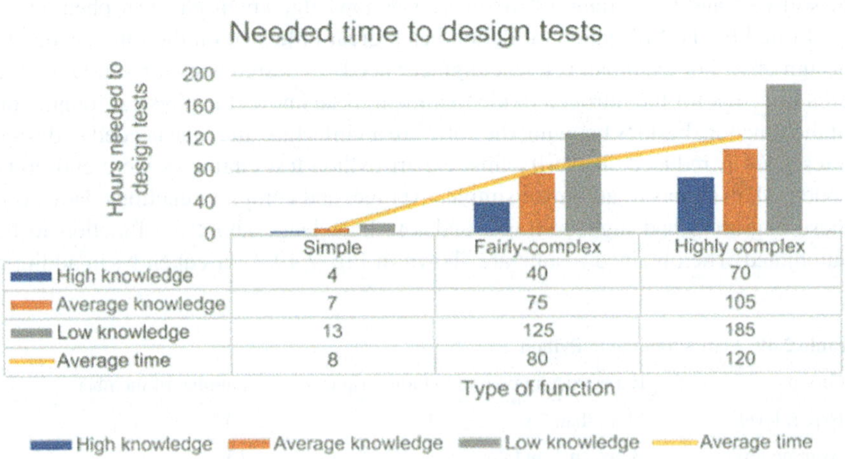

Fig. 2.40 Needed time for designing test cases vs functional and physical knowledge about a SM

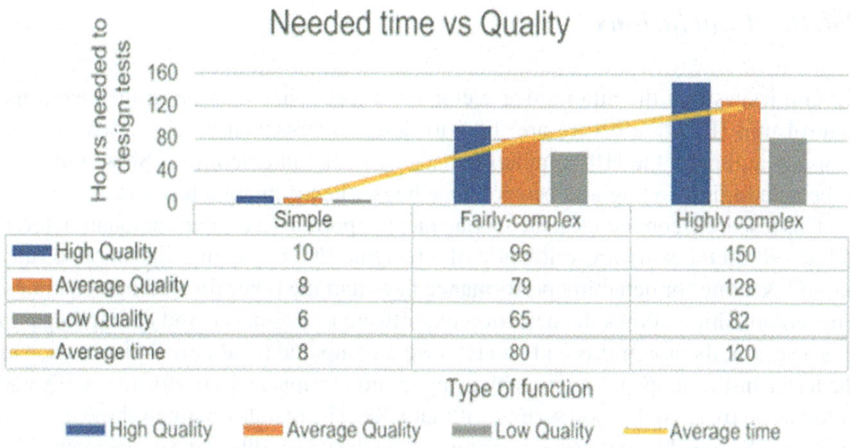

Fig. 2.41 Engineering hours spent for test design depending on the type of SM to be validated in black-method

shown in Fig. 2.42. Significant productivity improvements when comparing with the black-box technique can be obtained when the training of the staff is improved: 13.5% for complex functions, 10.9% for fairly complexity functions, and 16.6% for simple functions considering the average knowledge case. These figures are based on the scenario of high Python skills as well as good knowledge of the SM under validation.

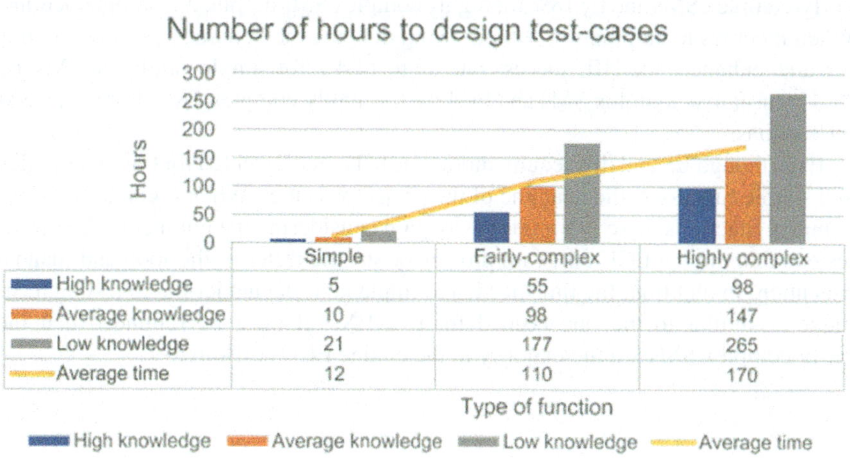

Fig. 2.42 Total number of hours to design the test cases (design and script coding time)

2.4.10 Conclusions

Several issues that the automotive sector must face when validating the electronic control unit (ECU) software are: how to design representative use-cases, how to properly automate the HIL simulation because of the interaction of SMs, and, how to be able to find coding and performance bugs when running a test-case.

This research, conducted at the second most important European car manufacturer, is focused on the software validation of an engine ECU by using dlls and two rule-based EXs, one for detecting performance bugs and the other for finding code bugs. This combination allows the detection of software performance and coding bugs. In this research, the use of dlls and two EXs were compared to other techniques such as the tester-in-the-loop, automation by using Python scripts and a performance EX and automation by using Python scripts without EXs. The results obtained show that dlls and two EXs are able to detect 6 bugs more than the use of dlls and a performance EX can, 14 bugs more than the tester-in-the-loop can, 16 bugs more than the automation by using Python scripts can,15 bugs more than a manual execution can and 14 bugs more than the model-based testing can. Dlls and EXs working in cooperation enhance the code coverage regarding the other techniques. This enhancement depends on the number of states in the functional model used in the EXs and the number of subintervals in which the SM inputs can be divided as shown in this research.

Dlls and Python scripts can be used combined with different techniques such as the using of a performance EXs or two EXs. The obtained results show that the methodology proposed in this research enhances the HIL success rate compared with the tester-in-the-loop technique by up to 6% for simple validation SMs, by 16.8% for fairly-complex SMs and by 18% for highly complex SMs despite the SM interactions. When it comes to automation without using dlls, the methodology proposed in this research enhances the HIL success rate up to 14.4% for simple validation SMs, by 27.4% for fairly-complex SMs and by 47% for highly complex SMs despite the SM interactions.

Even though ES and dlls require more time to be implemented for highly-complex and simple functions, the deadline of the project was met. When it comes to fairly-complex functions there is a productivity gain considering the number of SMs to be tested in an engine ECU software project versus the tester-in-the-loop and manual execution. In addition, the time needed to implement the model-based testing technique is similar to the one needed for two EXs. It must be reminded that the fairly-complex SMs are the majority in the engine ECU software.

References

1. Broy, M. Challenges in automotive software engineering. Proceedings of the 28th International Conference on Software Engineering (pp. 33–42). ACM. 2006
2. Navet, N., & Simonot, F. Automotive Embedded Systems Handbook (1st ed.). CRC Press. Florida, United States, 2008
3. Roychoudhury, A. Embedded Systems and Software Validation (1st ed.). Morgan Kaufmann Publishers, Massachusetts, United States, 2009
4. Gajjar, M. J. Sensor Validation and Hardware-Software Code Design. Mobile Sensors and Context-Aware Computing (1st ed.). Morgan Kaufmann, Massachusetts, United States, 2017
5. Rajan A., Wahl T. CESAR - Cost-Efficient Methods and Processes for Safety-Relevant Embedded Systems (1st ed.), Berlin, Germany, Springer, 2013
6. Oshana, R. Software Engineering for Embedded Systems: Methods, Practical Techniques and Applications (1st ed.), Amsterdam, Netherlands, Elsevier, 2013
7. Oberkampf, W. L., Roy, C. J. Verification and Validation in Scientific Computing (1st ed.). Cambridge University Press, Cambridge, United Kingdom, 2010
8. Westland, J. C. The cost of errors in software development: evidence from industry. *Journal of Systems and Software* 2002, 62(1), 1–9
9. Zaman, N. Automotive Electronics Design Fundamentals (1st ed.), Berlin, Germany, Springer, 2015
10. BOSCH, BOSCH Automotive Electrics and Automotive Electronics (1st ed.). Germany, Robert BOSCH, 2013
11. Garousi, V., & Mäntyla, M. V. A systematic literature reviews in software testing. *Information and Software Technology* 2016, 80, 195–216
12. Kasaju A., Petersen K., & Mantyla M. V. Analyzing an automotive testing process with evidence- based software engineering. *Information and Software Technology* 2013, 55, 1237–1259
13. Jungui Z., Zhiyi Z., Peizhang X., & Jingyu W. A test data generation approach for automotive software. Proceedings of the Conference IEEE 2015
14. Hoffmann A., Quantc J., & Wochrlc M. Experience report: White box test-case generation for automotive embedded software. Proceedings of the IEEE Ninth International Conference on Software Testing, 2016
15. Saglietti, F., Oster, N., & Pinte. F. White and grey-box verification approaches for safety and security critical software systems. *Information Security Technical Report* 2008, 13(1), 10–16
16. Wernick, P., & Lehman, M. Software process white box modelling for FEAST/1. *Journal of System and Software* 1999, 46(2–3), 193–201
17. Awedikian R., & Yannou, B. Design of a validation test process of an automotive software. *International Journal on Interactive and Manufacturing 2010*, 4(4), 259–268
18. Conrad, M. http://drops.dagstuhl.de/opus/volltexte/2005/325/, 2005
19. Chunduri, A. http://www.diva-portal.org/smash/get/diva2:945731/FULLTEXT02, 2005
20. Skruch, P., Buchala, G. Model-based real-time testing of embedded automotive systems. SAE *Int. J. Passeng. Cars – Electron. Electr. Sys* 2014. 7(2), 337–344
21. Raffaëlli, L., Vallée, F., Fayolle, G., Armines, A., Souza, P., Rouah, X., Pfeiffer, M., Géronimi, S., Pétrot, F., & Ahiad, S. Proceedings of the Embedded Real Time Software and Systems Conference, 2016
22. All4Tec, http://www.all4tec.net/MaTeLo/homematelo.html (last accessed on 10/03/2021)
23. Ilic, V., Popic, S., & Kovacic, M. Data flow in automated testing of the complex automotive electronic control units. *IEEE Instrumentation & Measurement Magazine* 2016
24. Keller, R., Alink, T., Pfeifer, C., Eckert, C. M., Clarkson, P. J., & Albers, A. Proceedings of the International Conference on Engineering Design, 2017
25. Köhl, S., Lemp. D., & Plöger, M. ECU network testing by hardware-in-the-loop simulation. *ATZ Worldwide* 2003, 105(10), 10–12
26. National Instrument, http://www.ni.com/white-study/10343/en/ (last accessed January 2020)

27. Winemantech, https://www.winemantech.com/services/hardware-in-the-loop-test-systems/ (last accessed January 2020)
28. Petrenko, A., Nguena, T., & Ramesh, S. Model-based testing of automotive software: some challenges and solutions. Proceedings of the 52th Congress ACM/IEEE Design Automation Conference, 2015
29. Matlabcentral File exchange. https://www.mathworks.com/matlabcentral/fileexchange/9709-from-simulink-to-dll-a-tutorial (last accessed January 2020)
30. NCA Software Product etas. https://www.Etas.Com/En/Products/Inca_Software_Products.Php (last accessed January 2019)
31. DSPACE. Simulator Hardware. https://www.Dspace.Com/En/Inc/Home/Products/Hw/Simulator_Hardware/Dspace_Simulator_Full_Size.Cfm (last accessed February 2022)
32. DSPACE Experiment and visualization. https://www.dSpace.com/en/inc/home/products/sw/experimentandvisualization/controldesk.cfm (last accessed February 2022)
33. DSPACE. Test Automation Software. https://www.dspace.com/en/pub/home/products/sw/test_automation_software/automdesk.cfm (last accessed February 2022)
34. Krûguer, M., Straube, S., Middendorf, A., Hahn, D., Dobs, T., Lang, K.D. Requirements for the application of ECUs in e-mobility originally qualified for gasoline cars. *Microelectronics Reliability* 2016, 64, 140–144
35. Gajjar, M.J. *Mobile Sensors and Context-Aware Computing*. Massachusetts: Morgan Kaufmann Publishers, Massachusetts, United States, 2017
36. Lockledge, J.C., Salustri, F.A. Defining the Engine Design Process. *Journal of Engineering design,* 10, 109–124. https://doi.org/10.1080/095448299261344 (last accessed March 2022)
37. Raikwar, S., Jijyabhau, L.W., Arun Kumar, S., Sreenivasulu Rao, M. Hardware-in-the-Loop test automation of embedded systems for agricultural tractors. *Measurement* 2019, 133: 271–280
38. Plummer, A.R. Model-in-the-loop testing, Proceedings of the Institution of Mechanical Engineers Part I *Journal of Systems and Control Engineering* 2006, 220 (3), 183–199
39. Zhan, Y., Clark, J.A. A search-based framework for automatic testing of MATLAB/Simulink models. *Journal of Systems and Software* 2008, 81(2), 262–285
40. Vivas, J.L., Agudo, I., Lopez, J. A methodology for security assurance-driven system development. *Requirements engineering* 2011, 16, 55–73
41. Martin, H., Ma, Z. , Schmittner, C. , Winkler, B., Kreiner, C. Combined automotive safety and security pattern engineering approach. *Reliability Engineering & System Safety* 2020, 198, Article 106773
42. Haghighatkhah, A., Banijamali, A., Pekka Pakanen, O., Oivo, M., Kuvaja, P. Automotive software engineering: A systematic mapping study. *Journal of Systems and Software* 2017, 128, 25–55
43. Hooshyar, H., Mahmood, F., Vanfretti, L., Baudette, M. (2015). Specification, implementation, and hardware-in-the-loop real-time simulation of an active distribution grid. *Sustainable Energy, Grids and Networks* 2015, 3, 36–51
44. National Instrument (2019). https://www.ni.com/fr-fr/innovations/white-studys/17/what-is-hardware-in-the-loop-.html (Accessed on 3 March 2020)
45. Ortega-Cabezas, P.M., Colmenar-Santos, A., Borge-Diez, D., Blanes-Peiró, J.J. Application of Rule-Based Expert Systems and Dynamic-Link Libraries to Enhance Hardware-in-The-Loop Simulation Results, *The Journal of Software 2019,* 14(6), 265–292
46. Meloa, S.M., Carver, J.C., Souza, P.S.L., Souza, S.R.S. Empirical research on concurrent software testing: A systematic mapping study. *Information and Software Technology,* 2019, 105, 226–251
47. Vandi, G. , Nicolò, C., Corti, E., Mancini, G. , Moro, D., Ponti, F., Ravaglioli, V. Development of a Software in the Loop Environment for Automotive Powertrain System. *Energy Procedia* 2014, 45, 789–798
48. Garousi, V., Felderer, M., Kilicaslan, F.N. (2009). A survey on software testability. Cornell University. https://arxiv.org/abs/1801.02201 (last accessed on 17 January 2020)
49. Walia, G.S., Carver, J. C. A systematic literature review to identify and classify software requirement errors. *Information and Software Technology* 2009, 51(7), 1087–1109

50. Ågren, S.M., Knauss, E., Heldal, R., Pelliccione, P., Malmqvist, G., Bodén, J. The impact of requirements on systems development speed: a multiple-case study in automotive, *Requirements Engineering* 2019, 24, 315–340
51. Dos Santos, J., Martins, L.E.G, de Santiago Junior, V.A, Povoa, L.V., dos Santos, L.B.R. Software requirements testing approaches: a systematic literature review, *Requirements Engineering* 2019, https://doi.org/10.1007/s00766-019-00325-w
52. Abadeh, M.N. Performance-driven software development: an incremental refinement approach for high-quality requirement engineering, *Requirements Engineering* 2020, 25, 95–113
53. Feldhütter, A., Segler, C., Bengler, K. Does Shifting Between Conditionally and Partially Automated Driving Lead to a Loss of Mode Awareness? In N. Stanton (Ed.), Advances in Human Aspects of Transportation. AHFE 2017. *Advances in Intelligent Systems and Computing* 2018, 597, pp. 730–741
54. ISO. Cybersecurity standard (2019). https://www.iso.org/standard/70939.html (last accessed on 20 September 2020)
55. Utesch, F., Brandies, A., Pekezou, P., Schiessl, F., Schiessl, F. Towards behaviour based testing to understand the black box of autonomous cars. *European Transport Research Review* 2020, 12, 48
56. Huang, W.L., Wang, K. Ly, Y., Zhu, F. Autonomous Vehicles Testing Methods Review. IEEE *19th International Conference on Intelligent Transportation Systems (ITSC)* 2016, pp. 163–168
57. Riedmaier, S., Ponn, T., Ludwig, B., Shick, F. Diermeyer, F. Survey on Scenario-Based Safety Assessment of Automated Vehicles. *IEEE Access* 2020, 8, 87456–87477
58. dSpace http://www.cokesen.com/resimler/1521204313_Dokuman1.pdf (last accessed 10 September 2020)
59. Möller, D., Haas, R. Guide to Automotive Connectivity and Cybersecurity. Berlin, Germany, Springer
60. El-Rewini, Z., Sadatsharan, K., Flor, D., Siby, S., Plathottam, J., Ranganathana, P. Cybersecurity challenges in vehicular communications. *Vehicular communications* 2020, 23, 100214
61. Vector. https://www.vector.com/int/en/know-how/technologies/safety-security/automotive-cybersecurity/#c2941(last (accessed on 10 September 2020)
62. Placho, T., Schmittner, C., Bonitz, A., Wana, O. Management of automotive software updates. *Microprocessors and microsystems* 2020, 78, 103257
63. Koegel, M., Wolf, M. (2018). Auto update – safe and secure over-the-air (SOTA) software update for advanced driving assistance systems. Berlin, Germany, Springer
64. ISO. Autonomous driving safety standard. https://www.iso.org/standard/70918.html (last accessed on 20 September 2020)
65. Banish, G. Engine Management: Advanced Tuning. Minnesota: Cartech, 2007
66. Garousi, V., Mäntylä, M.V. A systematic literature review of literature reviews in software testing. *Information and Software Technology* 2016, 80, 195–216
67. Kasoju, A., Petersen, K., Mäntylä, M.V. Analyzing an automotive testing process with evidence-based software engineering, *Information and Software Technology* 2013, 55(7), 1237–1259
68. Matelo® Software. https://www.all4tec.com/ (last accessed on 7 February 2020)
69. BOSCH. *BOSCH Automotive Electrics and Automotive Electronics* (1st ed.). Robert BOSCH. Germany. 2013
70. IEEE http://ieeexplore.ieee.org/document/4344112/ (last accessed June 2018)
71. Mariani L, Pezze M, Zuddas D. Recent advances in automatic black-box testing. *Adv. Comput.* 2015; 99: 157–193
72. Engström E, Runeson P, Skoglund M. A systematic review on regression test selection techniques. *Inf. Softw. Technol.* 2010; 52(1): 14–30
73. Linderman U, Maurer M, Braun T. Structural Complexity Management. 1st ed. Springer, Berlin, Germany. 2009
74. Yoo S, Harman M. Pareto efficient multi-objective test case selection. Proceedings of the ACM/ SIGSOFT. *International Symposium on Software Testing and Analysis* 2007. ACM. 2, 140-150
75. Zhou J., Zhang Z., Xie P., Wang J. A test data generation approach for automotive software. IEEE International Conference on Software Quality, Reliability and Security. 2015

76. Sopan-Barhate, Effective test strategy for testing automotive software. International Congress of Electronic Instrumentation and Control. 2015
77. Xing Y, Gong Y, Wang Y, Zhang X. Intelligent test-case generation based on branch and bound. *The Journal of China Universities of Posts and Telecommunications*. 2014; 21(2): 91-97
78. Zhang W, Yang Y, Wang Q. Using Bayesian regression and EM algorithm with missing handling for software effort prediction. *Inf. Softw. Technol.* 2015; 58: 58-70
79. Zheng J. Predicting software reliability with neural network ensembles. *Expert Systems with Applications,* 2009; 36(2, Part 1): 2116-2122. 29
80. Conrad, M. http://drops.dagstuhl.de/opus/volltexte/2005/325/ (last accessed January 2017)
81. Chunduri, A. (2016) http://www.diva-portal.org/smash/get/diva2:945731/FULLTEXT02 (last accessed January 2018)
82. Raffaëlli L., Vallée F., Fayolle G., Armines A, de Souza P., Rouah X., Pfeiffer M. Géronimi S. Pétrot F. Ahiad S. Embedded Real Time Software and Systems Conference. 2016
83. All4Tec. http://www.all4tec.net/MaTeLo/homematelo.html (last accessed November 2017)
84. Perez, Y.M., Marin, H.A.P., Bedoya, A.E. http://www.revistaieeela.pea.usp.br/issues/vol14issue5May2016/14TLA5_41EspinosaBedoya.pdf (last accessed January 2018)
85. Mechanical Simulation. https://carsim.com/products/realtime/index.php (last accessed January 2018)
86. National Instrument http://www.ni.com/white-study/10343/en/ (last accessed November 2017)
87. Petrenko A. Nguena-Timo, Ramesh S. Model-based testing of automotive software: some challenges and solutions. 52th Congress ACM/IEEE Design Automation Conference. 2015
88. Tatar M, Mauss J. Systematic Test and Validation of Complex Embedded Systems. Toulouse, France: Embedded Real Time Software and Systems (ERTS 2014); 2014

Chapter 3
Driver Efficiency and Software: Influence on Eco-Design

David Borge-Diez, Pedro-Miguel Ortega-Cabezas, Antonio Colmenar-Santos, and Jorge-Juan Blanes-Peiró

Abbreviations

API	Application Programming Interface
EC	Eco-charging
EDR	Eco-driving
ER	Eco-routing
EV	Electric Vehicle
GRU	Gated Recurrent Units
NAR	Nonlinear Autoregressive Neural Networks
RE	Renewable Energy
V2B	Vehicle-to-Building
V2G	Vehicle-to-Grid
V2H	Vehicle-to-Home
VCU	Vehicle Control Unit

Extract from *Energies* **2020**, *13*(15), 3997; https://doi.org/10.3390/en13153997, https://www.mdpi.com/journal/energies.

D. Borge-Diez · J.-J. Blanes-Peiró
Department of Electrical and Control Engineering, Universidad de León, Campus de Vegazana, s/n, 24071 León, Spain
e-mail: david.borge@unileon.es; dbord@unileon.es

P.-M. Ortega-Cabezas · A. Colmenar-Santos (✉)
Departamento de Ingeniería Eléctrica, Electrónica y de Control, UNED, Juan del Rosal, 12, Ciudad Universitaria, 28040 Madrid, Spain
e-mail: acolmenar@ieec.uned.es

P.-M. Ortega-Cabezas
e-mail: pedro-miguel.ortega-cabezas@valeo.com

© The Author(s), under exclusive license to Springer Nature Switzerland AG 2024
A. Colmenar-Santos et al. (eds.), *Development and Testing of Vehicle Software and its Influence on Sustainable Transport*, https://doi.org/10.1007/978-3-031-47630-3_3

3.1 Introduction

Designing eco-friendly products involves energy efficiency improvements. Eco-friendly products must consider not only raw materials and manufacturing processes to improve energy efficiency but also energy needed when designing them. This research shows how eco-routing (ER), eco-charging (EC), eco-driving (EDR), vehicle-to-grid (V2G) and electric vehicles (EVs) can contribute to the reduction of energy consumption during the product design. To do this, a group of 44 engineers assigned to the project was chosen to assess the total energy available for V2G when driving EVs from their homes to the design center by using ER, ED and, EC by running an application coded by the authors. The energy stored in EVs was used to quantify the reduction in energy consumption of the buildings present in the design center. The results show that energy saving ranges from 2.89% to 6.9% per day. In other words, 93 kWh per day during the design process. In addition, the fact of making the design process greener implies that renewable energies (REs) are integrated better during the design process. By running the application, drivers are informed about the RE mix when the charging process takes place. Finally, this research shows that current policies make V2G and vehicle-to-home (V2H) techniques not compatible.

Eco-design allows implementing eco-friendlier products as environmental impacts are considered during the design phase [1, 2]. Several factors which influence eco-design have been identified in several research [3–6]. Among them, one can find: manufacturing without producing hazardous waste, using clean technologies, reducing product chemical emissions and product energy consumption, using recycle materials and reusing components, designing products for ease of disassembly and reusing or recycling products at the end of their lives [7]. Some proposals have been made to improve these factors. Morgan and Liker [8] push to use lean manufacturing in design when developing products. As detailed by Rosen and Kishawy [9], this usage could imply that several alternatives used during the project can be assessed and, consequently, costs and benefits of eco-design can be set. When it comes to energy efficiency, research is mainly focused on final products and manufacturing processes. The former deals with the energy labelling concept which allows companies to create labels indicating the product energy efficiency [10]. The latter includes reducing energy consumption during the manufacturing process. Seow et al. describe a new outlook called '*Design for Energy Minimization*' aiming to provide transparency regarding to energy consumption during the manufacturing process to help inform design decisions [11]. Ka-Leung-Moon et al. propose guidelines for design and production of sustainable energy-saving fashion products [12]. RE integration plays an important role in eco-design. As detailed by Crul et al. [13], considering that the number of products that need electricity is increasing rapidly, it is of paramount importance to incorporate them in the design process. In their dissertation, they propose guidelines to integrate REs in the final product. Finally, waste is an important topic to be considered. For example, cybersecurity is a source of waste as detailed later. Tecchio et al. performed a detail analysis of the potentials of material efficiency to guarantee waste prevention and material reuse [14].

In addition to the manufacturing process, one can find many activities which generate pollutants during the product design. Among them, one can find: software validation processes, prototype product testing and building energy consumption [15–17]. Even emissions generated when engineers involved in projects commute to work impact eco-design. Software validation is an essential activity when designing automotive electronic control units (ECUs) [15, 18]. To do this, prototype vehicles and a considerable number of hours of hardware-in-the-loop simulation are needed [19, 20]. Consequently, pollution is caused. When designing combustion engines, activities such as engine tuning and driving tests cause pollution. However, engine design process lacks both guidelines and policies to limit emissions during the engine development contrary to the vehicles which have already been marketed and whose emission limits are clear and strict. Building efficiency is an essential topic which contributes to emission reduction as detailed in literature [21–23]. Finally, the vehicles used by the engineers who are involved in any projects are a source of pollution when commuting to their work. Some factors analyzed in this study such as the number of engineers participating in the project and the location of their homes regarding the design center may decrease or increase emissions.

ER, EDR and EC have an impact on energy efficiency when it comes to eco-design as analyzed in this research. EDR includes all driving habits which could reduce energy consumption and emissions. Nowadays, most cars are equipped with the system that informs how efficient the way of driving is. Qi et al. [24] investigated EDR by quantifying the energy potentially saved when applying to EVs. Sabrina et al. [25] also proposed a similar work in which continuous and on-demand feedback on driving behavior, and safety is conducted. Zhan et al. [26] discussed how systems in charge of monitoring the EV battery improve energy efficiency. ER helps the driver to find the most efficient route to go from point A to B considering several parameters such as real-time traffic conditions, road types and gradient, passengers' and cargo weight. Nunzio et al. [27] implemented a new model based on speed fluctuations and a road network infrastructure to set the best route. The University of California has worked on systems which are able to collect energy consumption data in-real world driving conditions with the aim of integrating them into eco route algorithms [28]. In this research, EC measures the contribution of RE when charging EVs showing the optimal moments to do this.

Energy consumption of buildings can be reduced by following several options. Haque and Raham describe in their study a comparison between solar photovoltaic mini-grid pumped hydroelectric storage versus battery storage [29]. Their main conclusion is that pumped storage is almost half as efficient yet more expensive than conventional battery storage. Another option to reduce energy consumption is the usage of buildings which integrate photovoltaic energy as described by Haque et al. [30]. V2G technology allows a better integration of REs and energy peak reduction [31, 32]. Nevertheless, V2G technology is completely influenced by policies and battery degradation as detailed by Uddin et al. [33]. Consequently, it is of paramount importance to analyze V2G from different social dimensions [34].

Finally, cybersecurity aims to protect ECUs from being violated by modifying the internal code and calibration. Of course, this could lead to critical situations

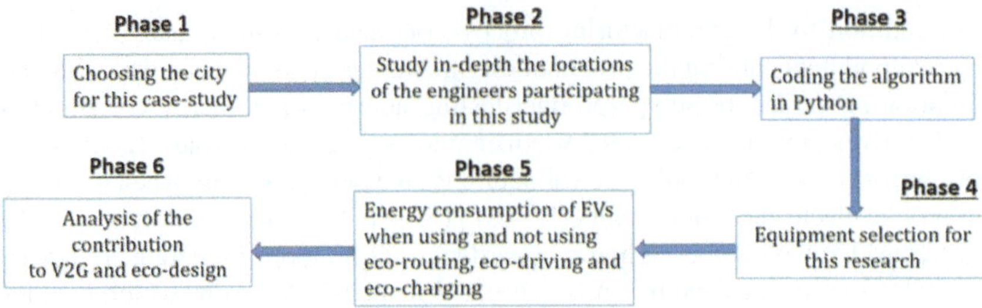

Fig. 3.1 Method followed in this research

where someone could take control of the vehicle. Several strategies can be followed. For example, a gateway can be integrated into the network architecture with the aim of keeping ECUs from being accessed in a reading or writing mode from an external computer unless this computer is connected to the manufacturer's network. Generally, an ECU stores several keys needed to assure its integrity. If an ECU fails, it must be analyzed whether there is a hardware or software problem. To do this, the security must be disabled. After that process, the ECU is not allowed to be installed in the vehicle again. As shown in this research, at that moment, the ECU is scrapped.

3.2 Methods

3.2.1 Overview of the Methods Used

When it comes to the method (Fig. 3.1), two key points must be considered to assess the contribution of ED, EDR and EC to eco-design. The former consists of choosing the optimal locations and the number of engineers participating in this research. The latter shows how the algorithm works and its implementation. To do this, the Here® API was used to determine the best route considering EC, EDR and ER models based on data from the vehicle control unit (VCU), traffic state, drivers' habits (the way of using the accelerator or brake pedals among other data), current battery capacity and potential recharge needs among others [35–37]. In addition, the EC concept shows drivers when a recharge process might be necessary taking into account when RE contribution is higher.

3.2.2 Description of Trips

Figure 3.2 and Table 3.1 depict the number of engineers participating in this study, the locations where these engineers live and the distance from each location to the design center. These locations were chosen to cover all parts of the city (North, South,

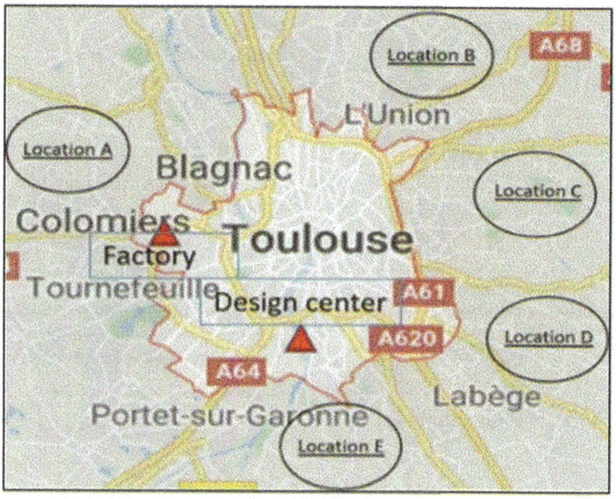

Fig. 3.2 Locations of engineers assigned to the project

Table 3.1 Distance between locations and the design center

From	To	Distance (km)	Number of engineers
Location A	Design center	24	10
Location B		20	8
Location C		26	8
Location D		22	12
Location E		19	6

East and West of Toulouse) and the number of engineers was assigned depending on traffic conditions in such a way that the more traffic jams are close to the location, the bigger number of engineers is chosen. All vehicles used were equipped with a 40-kWh battery. Their autonomy was close to 250 km. The maximum speed was 144 km/h, and the engine torque was around 320 Nm.

The trips were made in 2019 with the aim of collecting important data such as energy consumption by using the equipment.

3.2.3 *Algorithm Used in This Research*

The algorithm used in this research is depicted in Fig. 3.3. First of all, the energy consumption models available in the Here® API are tuned [36]. Then, the driver sets the destination by using a web interface. Afterwards, the algorithm assesses the optimal route for the driver. To do this, the Here® API is called by the Python code by using *Routingmode* parameter [35]. This parameter has an attribute named *Type* which can take three types of routes: the route that requires the least amount of travel time, the shortest one which reduces and optimizes the distance covered and finally,

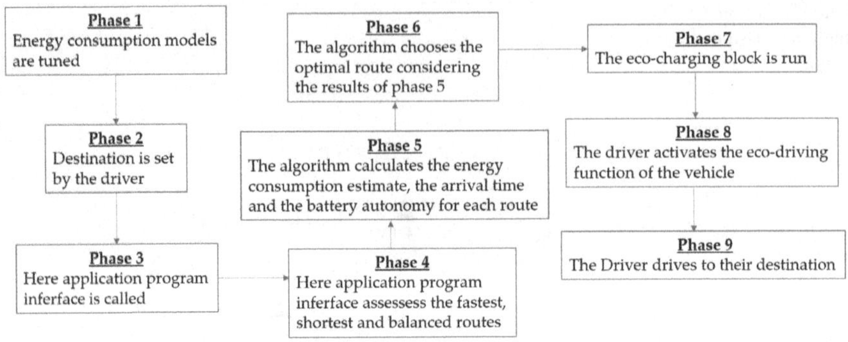

Fig. 3.3 Algorithm description

the balanced mode which searches for the correct balance between distance and time (only for trucks). The way how the algorithm works to determine the best routes belongs to the Here® know-how. The Python code receives from the Here® API the potential routes (the shortest, the fastest and the balanced one) to the destination and the energy consumption for each one. The application chooses the one with less energy consumption as explained later. Finally, the algorithm runs a block called EC which aims to calculate RE contribution and energy structure generation (wind power, photovoltaic, etc.) by using neural networks. The driver is, therefore, informed about when the charging process is greener.

Here® API provides energy consumption models that allow assessing energy consumption by using several parameters such as speed, auxiliary energy consumption (radio, cooling/heating, accelerations, decelerations, etc.). The way of tuning these models implies that the value of each parameter in kWh is provided depending on the speed value (if possible, as not all parameters are linked to speed such as auxiliary systems). In this research, these values were established by performing data acquisition after the drivers participating in this research made each trip 50 times in different periods and traffic conditions (Sect. 3.2.2). As shown in Fig. 3.4, Inca® software installed in a laptop as well as input/output from ETAS® supplier modules were used to perform these data acquisition. Finally, the tuning engineers of the company which collaborated in this study, assessed the factors values by analyzing data acquisition by using MDA® software and internal procedures. To introduce this information by using Here® interface is easy. First of all, the reader must indicate Here® that the standard energy consumption model will be used. Figure 3.5 shows an example that helps the reader to reproduce this study. Once these factors are tuned and introduced in the Python code, Here® returns the energy consumption estimate for each type of the routes (the fastest, the shortest and the balanced one). Consequently, the one with less energy consumption is chosen. Taking into account the initial battery capacity before the trip, the algorithm can determine if a charge is needed during the trip.

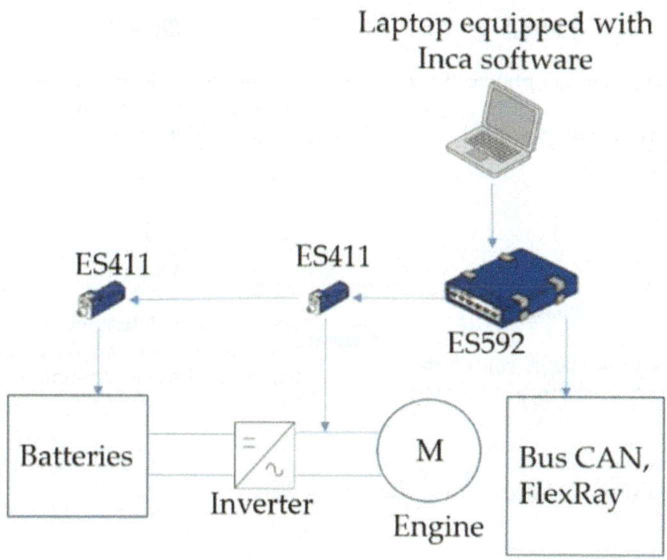

Fig. 3.4 Connection of the laptop to the EV

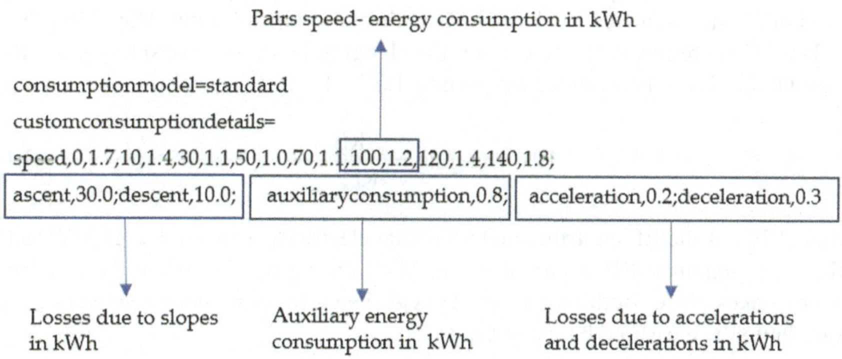

Fig. 3.5 How to configure the energy consumption model

Finally, the EC block is run, and the eco-score (how REs are integrated into the charging process) is assessed. The aim of this block is to determine the RE contribution when the charging process may take place considering the battery capacity. In addition, an estimate of energy structure (wind power, fuel, etc.) is made. The block is depicted in Fig. 3.6. In phase 1, several factors are analyzed such as the battery capacity and the energy consumption for a specific journey among others. It must be reminded that the energy consumption was estimated earlier by using the energy consumption model. Furthermore, the most likely time when the charging process takes place can be assessed (phase 2). Therefore, the RE contribution and most likely energy source mix (coal, solar energy, gas, etc.) can be obtained as detailed later by

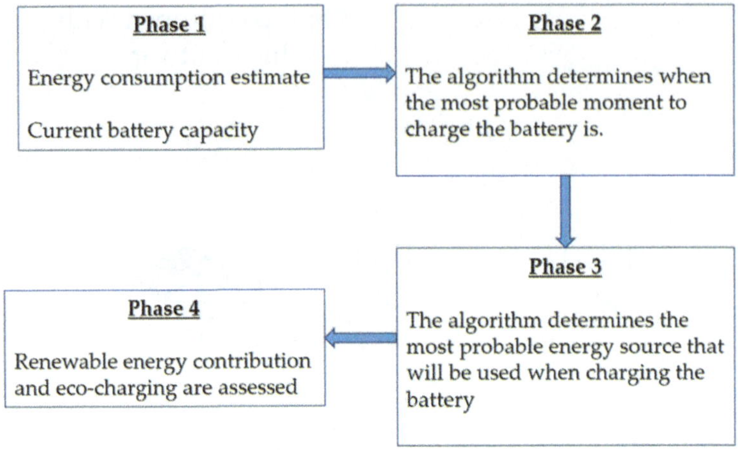

Fig. 3.6 Emissions estimated by using several parameters

using Gated Recurrent Unit (GRU) networks and nonlinear autoregressive (NAR) neural networks (phase 3) [38–42]. Finally, the EC is assessed considering the RE contribution. In addition, the algorithm proposed in this study provides information about different parameters such as chargers thanks to Open Charge Map API [35].

The EC score measures how green the charging process is considering the RE contribution. It can be assessed as given by Eq. (3.1):

$$\text{Eco-charging} = \frac{\text{RE}_{c,t}}{\text{RE}_{\max,d}} \tag{3.1}$$

where $\text{RE}_{c,t}$ is the RE contribution to the total electricity demand at t (in MW) and $\text{RE}_{\max,d}$ is maximal RE contribution (in MW) during the day when the charging process takes place. Both parameters are calculated by using neural networks. RE contribution is measured by using Eq. (3.2):

$$\text{RE}_c = \frac{\text{RE}}{\text{RE} + \text{NRE}} \tag{3.2}$$

where RE_c is the RE contribution (in %), RE is the total electricity generated by RE sources (in MW) and NRE is the total electricity generated by not RE such as coal (in MW).

$\text{RE}_{c,t}$ and $\text{RE}_{\max,d}$ are estimated as follows. The French system operator publishes files on a daily basis in which one can find the CO_2 generation structure and the total electricity demand of the day [43]. It must be taken into account that electricity demand and total RE contribution are stationary series. In other words, the pattern is repeated. Only some aspects have to be considered such as weekends and seasons. Anyway, two electricity consumption peaks can be found every day. Consequently, NAR networks are needed to model the electricity demand prediction for a specific day from a desired time (for example departure planned at 7 p.m.) to midnight. The Python code analyses the results returned by the neural network and determine the maximum RE contribution of the day. Finally, Eqs. (3.1) and (3.2) are assessed.

Typical recurrent networks present problems when it comes to long-term predictions due to the vanishing gradient problem. Engineers face this problem when training recurrent neural networks with gradient-based learning methods and back-propagation. When using this method, each of the neural network's weights receive an update proportional to the partial derivative of the error function with respect to the current weight in each iteration of training. In some cases, the gradient will be vanishingly small. Consequently, the weight does not change its value, and might stop the neural network training. To enhance long-term predictions, long-short term memory or GRU can be used. In this research, GRUs have been chosen, as they are more efficient (they require less memory). GRU is a recurrent neural network architecture that uses update and reset gates (Fig. 3.7).

Mathematically, the process is as follows:

a) Update gate for time step *t*

The update gate z_t is calculated by following Eq. (3.3):

$$z_t = \sigma \times \left(W^{(z)} \times x_t + U^{(z)} \times h_{t-1}\right) \tag{3.3}$$

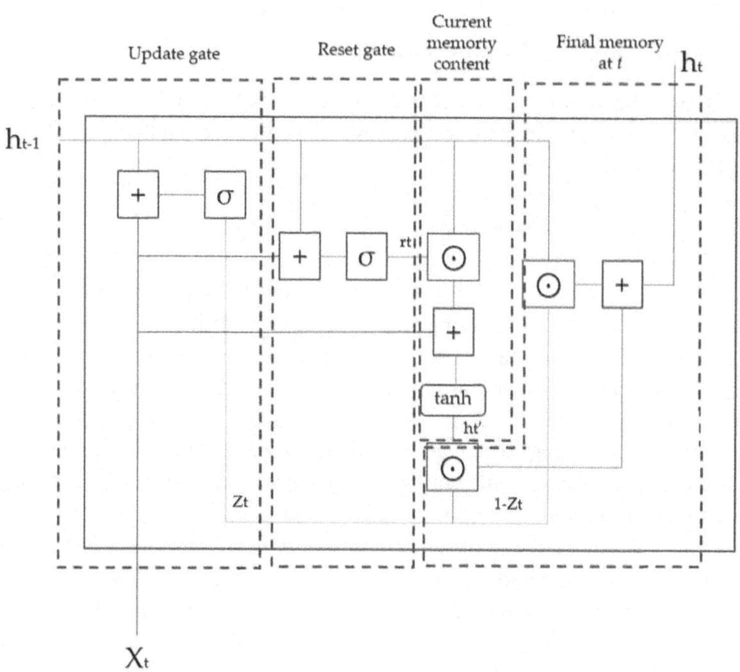

Fig. 3.7 GRU architecture

where x_t is the inputs presented to the network, $W^{(z)}$ is its weight matrix, h_{t-1} holds the information of the previous step $t-1$ and $U^{(z)}$ is its weight matrix. Both results are added, and a sigmoid activation is applied to squash the result between 1 and 0. The update gate allows determining how much of the past information should be passed along to the future.

b) Reset gate for time step *t*

It is given by Eq. (3.4).

$$r_t = \sigma \times \left(W^{(r)} \times x_t + U^{(r)} \times h_{t-1}\right) \tag{3.4}$$

The meaning of this factor is the same as for Eq. (3.3) unless r_t which is the reset gate. The reset gate corresponds to the past information which must be forgotten.

c) Current memory content

The new memory content h'_t uses the reset gate to store relevant information from the past (Eq. 3.5).

$$h'_t = \tanh(W \times x_t + r_t \times U \odot h_{t-1}) \tag{3.5}$$

The meaning of this factor is the same as for Eqs. (3.3) and (3.4). $\odot$ represents the Hadamard product.

d) Final memory at a current step

In this step the vector h_t is calculated by using Eq. (3.6). This vector holds the information for the current unit and passes it down to the network. To do this, the update gate is needed.

$$h_t = z_t \odot h_{t-1} + (1 - z_t) \odot h'_t \tag{3.6}$$

The GRU network was coded in Python. Figure 3.8 shows the pseudocode. To reproduce the results, the reader must have the data published by the French system operator for the last four years. The first three-year data are used for inputs of the network and the last-year data are employed as targets to train the network. It is of paramount importance to rescale all data to make them range between 0 and 1 to assure the network performance. The network parameters are set up by using the *keras* package. First of all, with the *Sequential* parameter, the code specifies that the model is sequential, and the output of each layer is the input for the next layer. In this study, the authors have used the *Dropout* function which is a technique where randomly selected neurons are ignored during training. This means that their contribution to the activation of downstream neurons is temporally removed on the forward pass and any weight updates are not applied to the neuron on the backward pass. The main advantage of this technique is that the network becomes less sensitive to the specific weights of neurons. The method used to analyze the error loss is the *mean squared error* which is widely recommended for regression problems. The method used to optimize the model is *Adam* which is an optimization algorithm that can be used instead of the classical stochastic gradient descent procedure to update network weights iterative based in training data. It offers many advantages such as straightforward implementation and computational efficiency among others. Other comments can be found in the pseudocode (Fig. 3.8).

The algorithm estimates the structure generation for the next two hours (Fig. 3.9) by using the data published by the French System Operator (CO_2 generation structure and the total electricity demand of the day) and NAR networks. These networks are useful when handling time series and predictions. These networks have been created and trained in an open loop. In this case, the targets are used as feedback. Then, the networks are verified in a close loop [38–40, 44]. Mathematically, NAR networks can be expressed by Eq. (3.7):

$$\hat{y}(t) = f(y(t-1) + y(t-2) + \cdots + y(t-d)) + \varepsilon(t) \tag{3.7}$$

```
#Import packages. Here some examples
Import numpy
Import pandas
import keras
import tensorflow
import sklearn

#import data published by the French System Operator
pd.read_csv(file_2016) # used as inputs
pd.read_csv(file_2017) # used as inputs
pd.read_csv(file_2018) # used as inputs
pd.read_csv(file_2019) # used as targets
#Prepare data to be used as inputs of the LSTM network
X=reshape_data_inputs_inputs # used as inputs
#rescale data to 0-1 scale
minimum = amin(X, axis=-1).reshape
maximum = np.amax(X, axis=-1).reshape()
X = (X-minimum) / (maximum-minimum)
Y = (Y-minimum) / (maximum-minimum)
#network parameters. A model is a stack of layers
model = Sequential()
#Adding layer with the number of inputs specified
model.add(GRU(128, input_shape=(data), return_sequences=True))
model.add(Dropout(0.1)) # Dropout = 10%
# Boolean. Whether to return the last output in the output sequence, or the full sequence.
model.add(GRU(64, return_sequences=True) #Adding layer with the number of inputs specified
model.add(Dropout(0.1)) # Dropout = 10%

# Boolean. Whether to return the last output in the output sequence, or the full sequence.
model.add(GRU(32, return_sequences=True)) ) #Adding layer with the number of inputs specified
model.add(Dropout(0.3)) # Dropout = 30%

#Optimizer choice and error measurement method
model.compile(loss='mean_squared_error', optimizer='adam')
#Training
his = model.fit(X, Y, batch_size=2, nb_epoch=5, verbose=1)# , callbacks=[TQDMNotebookCallback()])

#Plot error
plt.legend("mean squared error", loc="upper left")
```

Fig. 3.8 Pseudocode for the GRU networks

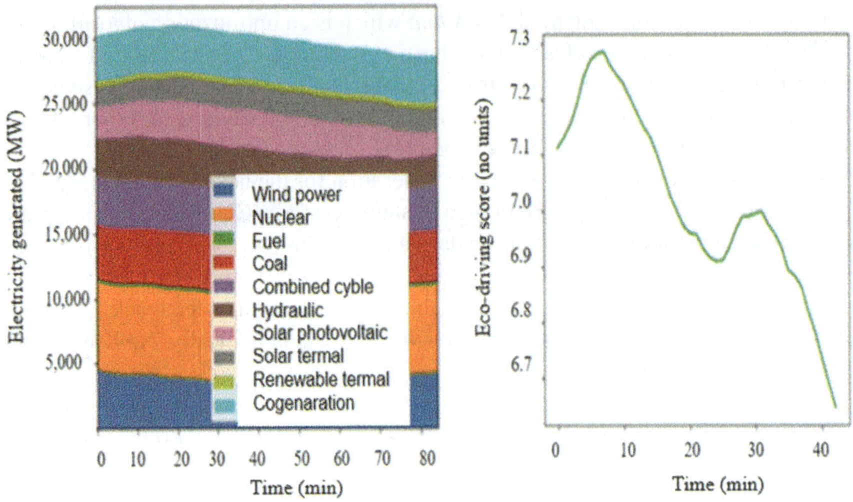

Fig. 3.9 Interface of the application: EDR and energy structure generation

where f represents the network response taking into account the previous input data, $\varepsilon(t)$ is the difference between the predicted value $\hat{y}(t)$ and the actual y. The number of delays establishes the d values to be considered for the prediction. The number of hidden layers and neurons per layer is flexible to achieve the best performance of the neural network under design. This number must be carefully chosen to avoid the increase of neural network complexity. The effect of choosing the value of the delay parameter is shown in Fig. 3.10. As one can see, a high d implies that predicted line series line changes slower. On the other hand, when d is lower, predicted line series follows the real power wind value more accurately. However, if d takes a very low value, then the predicted line series does not follow the real power wind value. The main explanation is that d determines the weigh given to past values. Consequently, significant changes in trend are not detected which could happen due to weather conditions. That is why, NAR networks are used in this research as an estimation and the accuracy remains on GRU networks. Anyway, this is not an issue as Matlab® allows correcting predictions if predicted values are known. This is the case of this application as it can predict $t+1, t+2, t+3\ldots$ at a specific moment t. However, when the moment is $t + 1$, the neural network can be updated as the predicted $t + 1$ value and the real $t + 1$ are known in real time (the French system operator publishes the needed data in real-time). To reproduce the results of this study, the authors obtained good predictions for the next 2 h with $d = 3$ when using the data belonging to 2019 published by the French system operator. The pseudocode of the NAR network is shown in Fig. 3.11 coded by using Matlab®. The NAR networks were trained by using *trainlm* function which implies that bias and weights are updated according to Levenberg–Marquardt optimization. It is the fastest backpropagation algorithm even if it may require more memory than other methods.

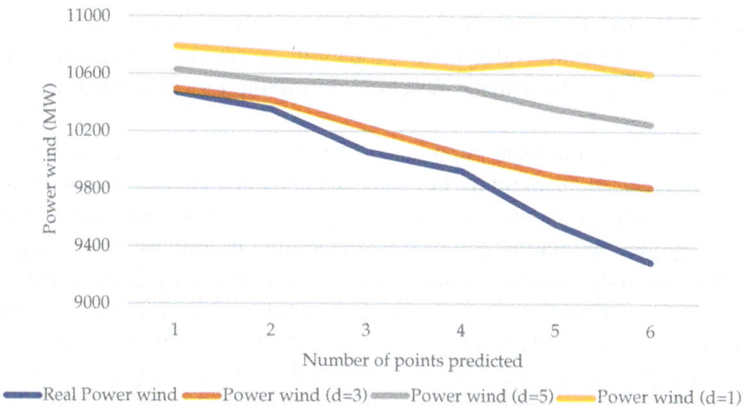

Fig. 3.10 Prediction vs real values depending on the delay parameter

```
%Pseudocode for Matlab®
data = French_system_operator_data; % loading data
net = narnet(1:3,10); % three delays and 10 hidden layer size. Train in open loop and training function
trainlm
[Xs, Xi, Ai, datas] = preparets(net,{},{},T); % preparing data to train
net = train(net,Xs,datas,Xi,Ai);  %train the network
[Y,Xf,Af] = net(Xs,Xi,Ai); %network performance assessment
perf = perform(net,datas,Y)
[netc,Xic,Aic] = closeloop(net,Xf,Af); %predicting results in close loop
```

Fig. 3.11 Pseudocode for NAR networks

3.2.4 Data Analysis

As detailed in the result section, the data obtained in this research seem to be close to a normal distribution. Consequently, a method must be set to confirm this assumption. To do this, the package named *PASSWR* belonging to R software was used. This package includes commands such as *EDA* which provide a lot of information to perform exploratory data analysis such as Kurtosis, Skewness and *p*-value. Kurtosis is a statistical measure that defines how heavily the tails of distribution differ from the tails of a normal distribution. Therefore, kurtosis identifies whether the tails of a given distribution contain extreme values. For a normal distribution, its value is 3. There are three types of kurtosis. Mesokurtic when kurtosis is close to 3. Leptokurtic when values are quite higher than 3. Platykurtic when the extreme values are less than the normal distribution. Skewness essentially measures the symmetry of the distribution. For a normal distribution, its value should be close to 0. At this point, it is important to highlight that symmetry does not imply that the data correspond to a normal distribution. Thus, these two parameters must be analyzed carefully. Finally, the *p*-value or probability value is the probability of obtaining test results at least as extreme as the results actually observed during the test, assuming that the null hypothesis is correct.

Plots are also of paramount importance when analyzing the data. In this research, three plots were used: histograms, Q-Q plots and boxplot. A histogram is a graphical representation which organizes a group data points into user-specified ranges. The Q-Q plot, or quartile-quartile plot, is a graphical tool used to assess if a set of data plausibly came from some theorical distribution such as a Normal one. Finally, a box plot is a graphical rendition of statistical data based on the minimum, first quartile, median, third quartile, and maximum. In this graph, the top of the rectangle indicates the third quartile, a horizontal line near the middle of the rectangle indicates the median, and the bottom of the rectangle indicates the first quartile.

Figure 3.12 shows an example of how a dataset corresponds to a normal distribution by using *PASSWR*.

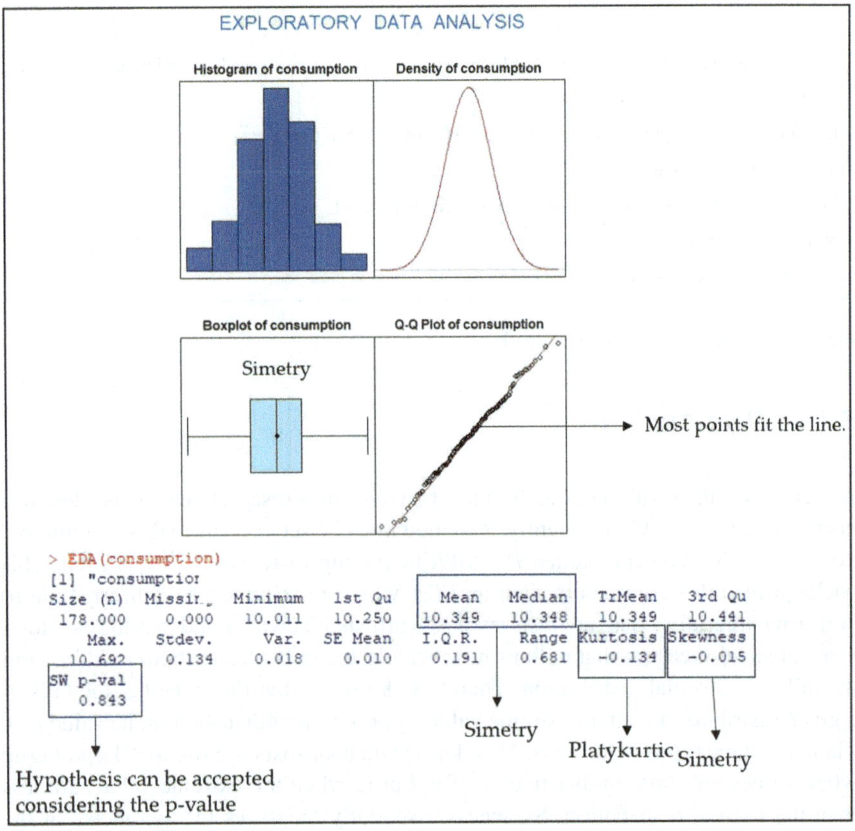

Fig. 3.12 Example of the exploratory analysis performed in this research

3.2.5 V2G

The energy consumption of 5 buildings located in the design center and the contribution of V2G to reduce their energy consumption during the design product were analyzed. Among the means that consume electricity in these buildings, one can find those that consume a great amount of energy such as test benches and, on the contrary, a considerable number of others such as computers, faxes, photocopying machines which have much lower energy consumption. It is of paramount importance to assess the amount of energy that EV can inject to reduce emissions and energy consumption during the design product.

3.3 Results

3.3.1 Distribution of Engineers

The project considered in this research deals with designing an ECU. As shown in Fig. 3.2 and Table 3.1, the reader can find the city and the location of all engineers participating in this project. Table 3.2 shows the statistical results achieved when processing the data obtained during the trips with no traffic jams. Without any traffic jams, the time needed to go from each location to the design center is very similar. Skewness is close to zero. Consequently, the distribution is symmetric. Kurtosis values show that the data distribution tails do not differ from normal distribution ones. The *p*-value represents the null hypothesis: the data follow a normal distribution. The null hypothesis can be considered as True if p-value > 0.05. Considering that the Skewness and the Kurtosis are sensitive to the sample size, the normality test was also confirmed by using Q-Q plots and histogram, which confirmed the normal hypothesis.

Table 3.2 Statistics parameters obtained without any traffic jams

Factor	Location A		Location B		Location C		Location D		Location E	
	N.A.[a]	A.U.[b]	N.A.[a]	A.U.[b]	N.A.[a]	A.U.[b]	N.A.[a]	A.U.[b]	N.A.[a]	A.U.[b]
Mean	8.9	8.7	7.8	7.5	9.9	9.6	8.4	8.1	7.2	6.9
Std deviation	0.2	0.15	0.18	0.15	0.16	0.12	0.19	0.15	0.16	0.12
Kurtosis	3.9	4.1	3.7	4.5	4.1	4.2	4.1	3.9	4.1	4.3
Skewness	−0.135	−0.121	−0.041	−0.032	−0.025	−0.015	0.035	0.03	0.08	0.045
p-value	0.412	0.452	0.425	0.454	0.325	0.343	0.385	0.410	0.396	0.332

[a]N.A. means no algorithm is used
[b]A.U. means the algorithm is used

Table 3.3 Statistics parameters obtained with traffic jams

Factor	Location A		Location B		Location C		Location D		Location E	
	N.A.[a]	A.U.[b]	N.A.[a]	A.U.[b]	N.A.[a]	A.U.[b]	N.A.[a]	A.U.[b]	N.A.[a]	A.U.[b]
Mean	12.5	10.3	10.8	8.7	13.2	11.5	11.2	9.4	11.5	8.3
Std deviation	0.9	0.89	0.98	0.95	0.75	0.81	0.99	0.88	1.15	1.01
Kurtosis	3.2	3.4	3.15	3.35	3.25	3.45	3.185	3.36	3.15	3.58
Skewness	0.281	0.112	0.231	0.189	0.261	0.185	0.259	0.189	0.262	0.215
p-value	0.356	0.411	0.389	0.422	0.321	0.468	0.369	0.498	0.311	0.336

[a]N.A. means no algorithm is used
[b]A.U. means the algorithm is used

Table 3.4 Statistics parameters obtained in mixed traffic conditions

Factor	Location A		Location B		Location C		Location D		Location E	
	N.A.[a]	A.U.[b]	N.A.[a]	A.U.[b]	N.A.[a]	A.U.[b]	N.A.[a]	A.U.[b]	N.A.[a]	A.U.[b]
Mean	10.8	9.2	9.5	8.1	12.1	10.5	10.6	8.8	10.1	7.5
Std deviation	0.6	0.5	0.9	0.6	0.75	0.65	0.85	0.5	0.65	0.58
Kurtosis	3.5	3.7	3.45	3.98	3.8	3.9	3.68	3.56	3.98	3.99
Skewness	0.145	−0.081	0.158	0.148	0.225	0.118	2.435	0.195	2.72	0.238
p-value	0.359	0.401	0.256	0.385	0.458	0.453	0.399	0.401	0.358	0.367

[a]N.A. means no algorithm is used
[b]A.U. means the algorithm is used

Tables 3.3 and 3.4 show the statistical results achieved when processing the data obtained during the trips with traffic jams and mixed conditions. The way of interpreting the results is similar to Table 3.2. Q-Q plots and histogram confirm the assumptions.

3.3.2 Building Energy Consumption

The design center is composed of 5 buildings with different configurations. Table 3.5 shows the characteristics of each building. These items were chosen after having analyzed the energy consumption data provided by the company participating in this research. Only the most highly energy consuming items were considered.

Figure 3.13 depicts the energy consumption for each building. These values were obtained by using electric meters installed in the design center, the number of hours of operation considering timetables and average energy consumption of facilities/items. The main differences between these buildings were the number of people

Table 3.5 Characteristics of buildings

	Building 1	Building 2	Building 3	Building 4	Building 5
Number of floors	1	1	1	2	2
Number of meeting rooms	2	4	3	8	12
Number of offices	10	12	10	20	40
Type of the heating systems	Electric	Electric	Electric	Electric	Electric
Type of the cooling systems	Electric	Electric	Electric	Electric	Electric
Lighting	Fluorescent low power energy consumption lighting system	Fluorescent low power energy consumption lighting system	Fluorescent low power energy consumption lighting system	Fluorescent low power energy consumption lighting system	Fluorescent low power energy consumption lighting system
Number of people in the building	20	25	20	40	80

working, which affects the number of other elements present in the buildings such as printers, and the presence of laboratories.

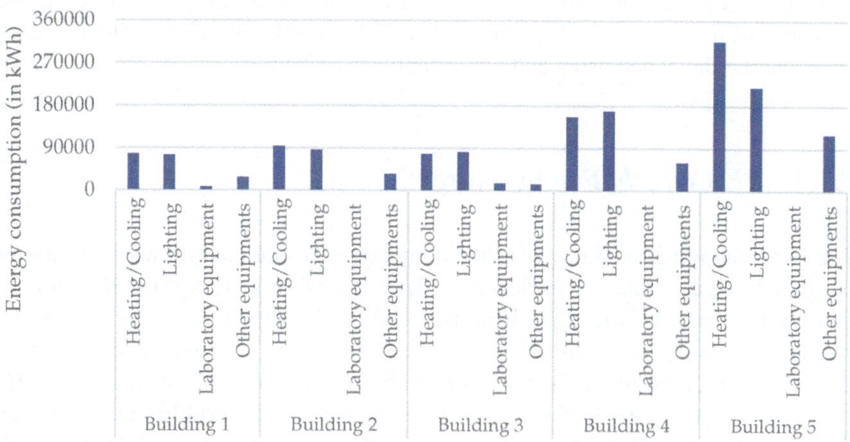

Fig. 3.13 Energy consumption of each building

Table 3.6 Summary of energy saving

	Total energy consumption (in kWh)	Total energy available when using the algorithm per day in kWh	Total energy available without using the algorithm per day in kWh	Contribution to meet energy consumption when using the algorithm (in %)	Contribution to meet energy consumption without using the algorithm (in %)	Delta per day without using the algorithm (in kWh)	Delta per year when using the algorithm (in kWh)
Without traffic jams	7,540.1	1,397.6	1,384.4	18.5	18.4	13.2	2,904
With traffic jams	7,540.1	1,332.8	1,239.6	17.7	16.4	93.2	20,504
Mix with and without traffic jams	7,540.1	1,376.9	1,337.1	18.3	17.7	39.8	8,756

Considering the results shown in Tables 3.2, 3.3 and 3.4, the energy to be used for V2G technology is assessed. However, V2G must be also compatible with V2H technology. Logically, the drivers will be willing to save this available energy for their own homes rather than injecting it into the grid. This topic will be discussed in the result discussion Sect. 3.4.3.

Table 3.6 depicts the main results obtained in this research. An additional gain of 2.89% to V2G can be obtained when considering mixed traffic conditions (with and without traffic jams). This percentage can be increased by 6.9% considering the influence of traffic jams. This percentage was obtained by dividing the total energy available per day when using and not using the algorithm.

3.3.3 RE Contribution to Charge EVs

The EVs belonging to employees should inject power into the grid when electricity consumption peaks take place. Generally, every day there are two peaks of electricity consumption which depend on the season and on the day of the week (Fig. 3.14). Anyway, these peaks usually occur between 8 a.m. and 1:30 p.m. and from 5:30 p.m. to 8 p.m. In this research, power should be injected to reduce the first electricity consumption peak. The fact of injecting this power into the grid from EVs implies that EVs should be recharged in some cases by the user to have enough energy to come back home. Consequently, it is of paramount importance to charge them when REs are being used. In other words, the recharge process should be done when the mix is greener. The algorithm proposed by the authors allows determining when the RE contribution is higher by using Eqs. (3.1) and (3.2). Therefore, the charging process is greener.

Table 3.7 shows the electricity prices offered by different suppliers in France.

Considering the prices shown in Table 3.7 and Fig. 3.14, several conclusions can be drawn: Firstly, when RE contribution is high, charging the EV battery is expensive, and the other way around. Secondly, drivers participating in V2G and V2H should charge their EVs between off-peak periods. Consequently, rapid or fast charging should be used. Consequently, battery life is reduced. Finally, the RE contribution vs traditional energy sources during off-peak periods must be increased.

Policies should change to increase RE contribution when charging EVs especially during off-peak periods. The French Government announced a new plan to increase the usage of RE. This plan has new different objectives for 2023. In this new plan, two scenarios are unfolded: an optimistic and a pessimistic one. The most important

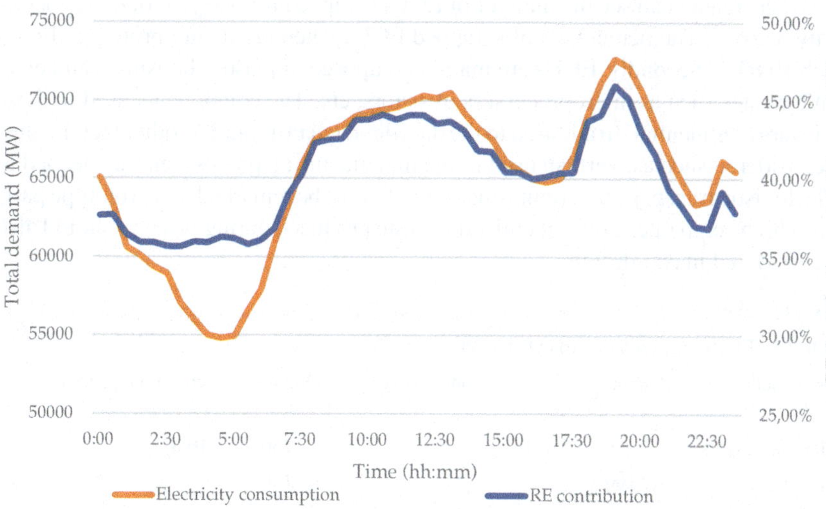

Fig. 3.14 RE contribution for January 2018. *Source* French system operator

Table 3.7 Electricity prices

Supplier	Price for off-peak periods (€/kWh)	Price for peak-periods (€/kWh)
Supplier 1	0.1230	0.1580
Supplier 2	0.1272	0.1638
Supplier 3	0.1280	0.1660
Supplier 4	0.1161	0.1483
Supplier 5	0.1138	0.1453
Supplier 6	0.1180	0.1513

figures imply that the total power installed should reach 69,980 MW in the worst-case scenario and 76,743 MW in the optimistic one. In addition to this, 150 and 167 TWh renewably sourced electricity should be delivered respectively. It must be reminded that the current renewable capacity in France is close to 56,000 MW. Some important figures to be retained [45]:

- Primary energy consumption by RE: 10.7%.
- Gross electricity production by ER in 2017: 92.6 TWh.

3.3.4 Cybersecurity

As aforementioned in the previous sections, cybersecurity could involve a considerable amount of waste to be considered. Table 3.8 shows the results of this research considering 120,000 ECUs in series production and 1,200 prototype units per year. Firstly, the reader can see the number of ECUs scrapped for one year due to cybersecurity reasons. The percentage of scrapped ECUs when producing prototype ECUs reach 10.83%. Secondly, ECUs are mainly composed of plastic (housings) and electronic wastes such as microprocessors, resistors, etc. The components used in these ECUs are confidential. In addition to the data depicted in Table 3.8, other factors must be considered such as: varnish used in the manufacturing process, energy needed to weld, logistic transportation (emissions), etc. It must be remarked that prototype parts have a high scrap rate. Consequently, the waste produced during projects should also be considered in eco-design.

Table 3.8 ECUs scrapped due to cybersecurity

Component	Quantity	Series production (%)	Quantity	Prototype production (%)
Microprocessors	190	0.16	130	10.83
Printed circuit board	190		130	
Housing	380		260	
Capacitors	17,100		11,700	
Resistors	21,100		14,436	
Programmable components	190		130	
Input/output connectors	190		130	
Memory RAM	280		260	

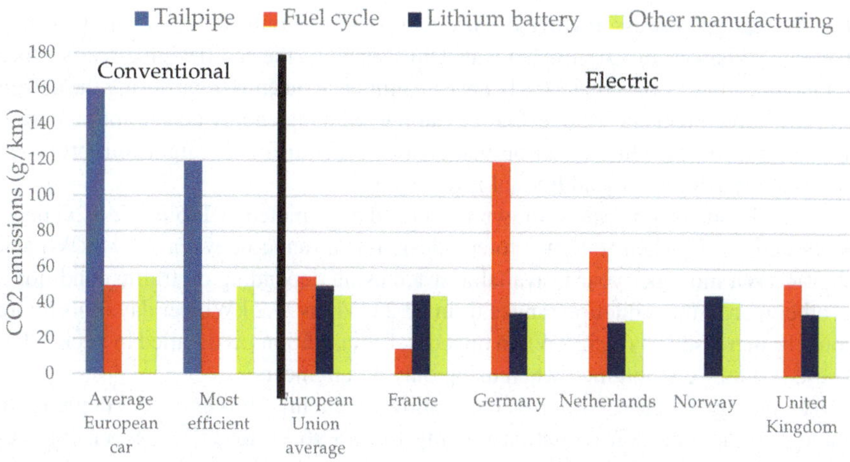

Fig. 3.15 Life-cycle emissions from electric and conventional vehicles. *Source* The International Council on Clean Transportation

3.4 Discussion

3.4.1 Sustainability of This Solution

No matter what kind of powertrain is used in a vehicle (hybrid, electric, diesel or gasoline), CO_2 emissions are always present as analyzed by the International Council on Clean Transportation [46]. As shown in Fig. 3.15, EVs pollute less than conventional vehicles when it comes to life-cycle emissions. Therefore, the algorithm and the solution proposed in this research are sustainable despite the process of charging and discharging that harms the battery. In addition, the EVs increase is confirmed by such a supplier as Continental®, considering the latest news concerning the recent sale of its diesel and petrol powertrain unit. Thermic engine activities are supposed to be reduced while EV activities are most likely to be increased in the year to come.

3.4.2 Energy Efficiency Improvement

Several policies are pursued with the aim of improving energy consumption based on increasing the RE usage, eco-design and building energy consumption. When it comes to eco-design, energy efficiency is a key factor and engineers try to improve energy consumption during the manufacturing process. When it comes to design processes, the eco-design process is mainly focused on the usage of recyclable raw materials and on the design of products that do not pollute much during their lives and they can also be recycled easily at the end of their lives. This research shows that the design process involves many factors that should be considered as they generate emissions in a significant way. Unfortunately, the eco-design directive does not include many topics that clearly have an influence on emissions during the design process. Investments in V2G technology allow reducing energy consumption during the design process. This reduction is increased when using the algorithm proposed in this study without any additional investment.

Several conclusions can be drawn from the date depicted in Table 3.5. V2G plays an essential role when it comes to eco-design as a range between 2,904 kWh and 20,504 kWh more per year is available when using depending on traffic conditions. In other words, the reduction ranges from 13.2 kWh to 93.2 kWh per day. This figure could be increased by 61.52 kWh in mixed traffic conditions as detailed in Sect. 3.4.4. These results were obtained when using only 44 engineers.

The energy efficiency improvement allows reducing emissions. As depicted in Table 3.9, the emission reduction ranging from 8.96 kg to 23.55 kg per day was obtained when using the algorithm proposed in this study. These values were obtained taking into account the energy saving (in kWh) and the monthly emission average from January to July 2019. It must be remarked that only 44 engineers participated in this study.

Finally, the fact of reducing energy consumption by using V2G is a topic which has been subjected to research. The contribution may be significant when the smart grid will be fully deployed. However, it should be also considered that V2G must be compatible with V2H. As shown in this research, it is not always the case. Several axes are essential to make them work together. The first one is improving battery capacities in such a way that the owner can use the energy stored in the battery for both purposes (V2H and V2G). The second one is also linked to batteries as the fact of going through charging and discharging processes should not degrade batteries quickly. Finally, RE contribution must be increased as described in Sect. 3.3.3.

Table 3.9 Emissions of CO_2 (in g/kWh) on a monthly basis

	January	February	March	April	May	June	July
Emissions (g/kWh)	55.68	51.30	31.24	26.32	24.74	27.65	35.85
Emission reduction per day (kg/kWh). Mixed traffic	2.22	2.04	1.24	1.05	0.98	1.1	1.43
Emission reduction per day (kg/kWh). No traffic jams	5.19	4.78	2.91	2.45	2.31	2.57	3.34

Source French System Operator

3.4.3 Renewable Energies

From a theoretical point of view, it is vital to consider V2G technology to reduce energy consumption of design buildings. In this case, two factors must be discussed. The first one is EV penetration into the market. The second one is policies. The former is essential due to the fact that the more EVs are sold, the more energy is available for V2G. Policies try to encourage drivers to choose EVs instead of traditional powertrains such as gasoline or diesel engines. Among these measures, one can find exemption from vehicle registration duties or municipal tax discounts. Despite this, the participation in V2G remains unclear for several reasons:

Drivers who cover many kilometers on a daily basis will be forced to choose between V2G and V2H techniques. Considering that if drivers inject energy during the day to reduce the energy consumption in buildings, they will probably need to charge as soon as they get to their homes. Consequently, they cannot participate in V2H.

The fact of reducing prices to make drivers charge their vehicles after 2 p.m. is still far from being the solution as RE contribution is not substantial. Therefore, it is essential to promote RE facilities.

The algorithm proposed in this research allows choosing the best moment to charge the EV battery considering the electricity mix. Considering all the aforementioned facts, V2G can contribute in a significant way to the reduction of emissions during the design process and should be also a key element to be considered in eco-design. However, the percentage of improvement is completely linked to policies associated with EV recharging.

3.4.4 Population Participating in This Study

Several factors must be considered when it comes to the population participating in this research. Firstly, this research was conducted considering a small size research and development center. It must be taken into account that there are much bigger design centers in France such as the one of Renault located in Lardy (1,600 workers) or the one of PSA Peugeot-Citroen located in Carrières-sous-Poissy (1,300 workers). Therefore, the number of engineers who can participate when implementing the algorithm will increase. However, policies are still important to encourage people to participate in V2G [33, 47].

Figure 3.16 shows an estimate when choosing a greater number of engineers (400 engineers) and the energy consumption calculated in Sect. 3.3.1, and the assumption shown in Table 3.10. When it comes to the number of engineers, the values were assigned by respecting the percentage of engineers of each location established for the test-case. When it comes to the number of times with and without traffic jams, these values were considered for the whole year. In addition, an error estimate has

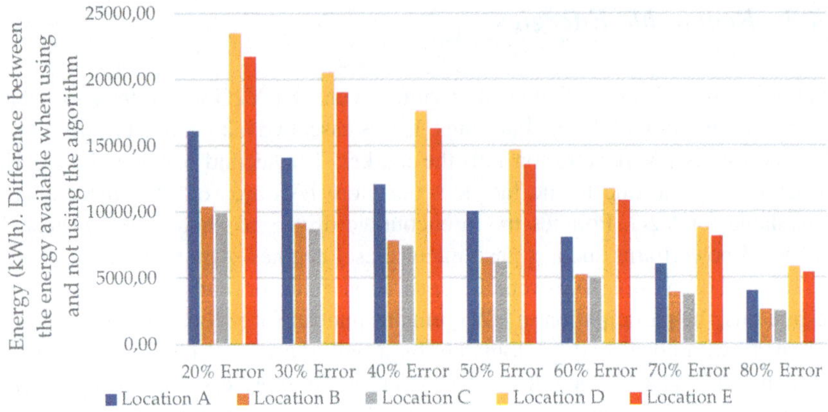

Fig. 3.16 Energy available depending on the error in energy consumption estimate when using 400 engineers

been added, which represents the nominal estimate with a 20% error, 30% error and so on (Table 3.11).

Table 3.10 Assumptions when considering a greater number of engineers

	Number of engineers	Number of times without and with traffic jams
Location A	91	300/60
Location B	73	320/40
Location C	73	316/44
Location D	109	180/120
Location E	54	170/140

Table 3.11 Gain when considering a greater number of engineers

	20% error	30% error	40% error	50% error	60% error	70% error	80% error
Energy available per year	81,824	71,596	61,368	51,140	40,912	30,684	20,456
Energy available per day	371.93	325.44	278.95	232.45	185.96	139.47	92.98
Gain per day in € (0.1347 €/kWh)	50.09	43.84	37.57	31.31	25.05	18.79	12.52
Gain per year (220 working days)	11,021	9,643.98	8,266.26	6,888.55	5,510.84	4,133.13	2,755.42

Figure 3.16 represents the increase of energy available for V2G when using the algorithm. Location D and Location E overperform the rest of the location when it comes to energy savings. The number of engineers (in other words the number of EVs) is not only the most relevant parameter to obtain more or less contribution to V2G and eco-design as Location E is the one which has fewer engineers/EVs participating in this case-study. Consequently, location is an important parameter.

Another important topic to be discussed is that this algorithm does not require investing in new facilities. Once the design center is adapted to use V2G, no more action is needed except for using this algorithm. As shown in Table 3.11, the fact of using it implies that an extra gain by 11,021 € could be obtained in this authors' estimate. Table 3.11 also shows the gain (difference between the energy available when using and not using the algorithm).

When considering the number of engineers participating in this case-study, the results are similar (Fig. 3.17). Again, location D and E account for the biggest amount of energy available for V2G. Consequently, the V2G energy value does not only depend on the number of EVs (engineers) participating in this research but the location of the engineers.

Choosing the optimal location is of paramount importance. If 22 EVs had been assigned to location D and E respectively (Fig. 3.18), the energy available for V2G will be increased from 39.8 kWh to 61.52 kWh per day.

Several factors such as distance and traffic state among others determine the optimal location to optimize the V2G contribution. Therefore, models which can predict the V2G contribution when determining the optimal number of engineers and location must be developed, as the fact of using more EVs does not imply that the contribution to V2G has increased in a significant way. This algorithm can be used off-line to determine the optimal locations by following the procedure shown in Fig. 3.19 Here® API can assess the routes off-line to obtain the time needed and,

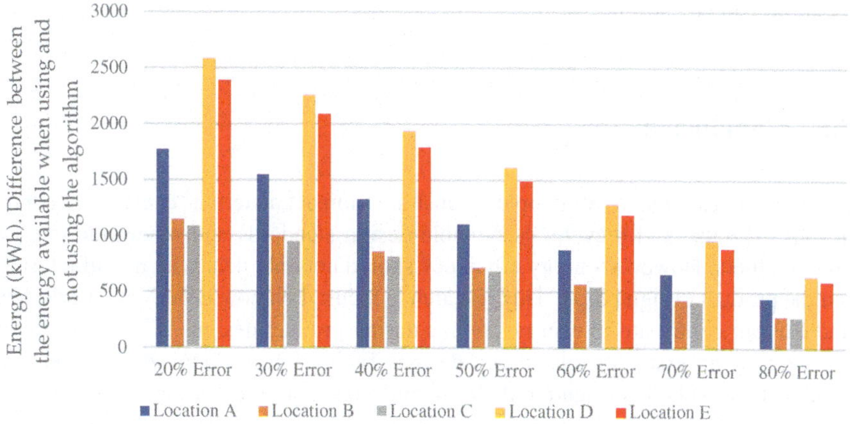

Fig. 3.17 Energy available depending on the error in energy consumption estimate when using 44 engineers

Fig. 3.18 Energy available depending on the error in energy consumption estimate when using 44 engineers only for destinations D and E

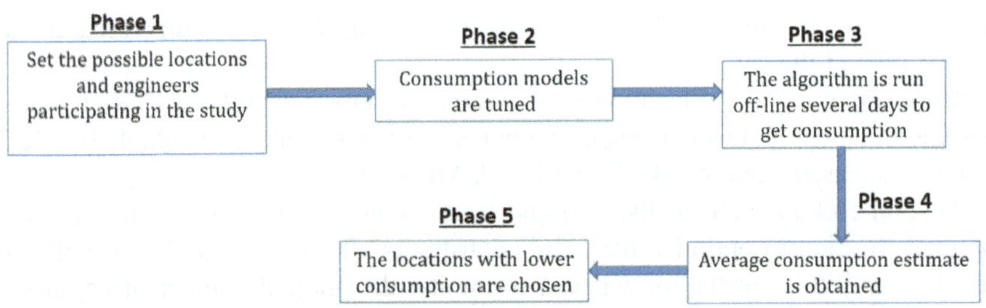

Fig. 3.19 Method to choose the optimal locations

consequently, the EV energy consumption. Thanks to this procedure, the energy consumption estimate for each driver can be obtained and, consequently, the optimal locations can be chosen.

3.5 Conclusions

Eco-design deals with several topics such as low-impact materials, energy efficiency, design for reuse and recycle, sustainable design standards and renewable energy among others. However, energy efficiency should not only deal with manufacturing but the product design phase. This research is focused on how EVs, V2G, EDR, ER and EC can contribute to energy savings during product designs. Therefore, these factors play an essential role in eco-design. Taking into account the method and results obtained in this research, the following conclusions are drawn:

a. Energy savings

The algorithm proposed in this research which uses energy consumption models properly tuned for EDR, ER and EC allows reducing energy consumption between 2.89% and 6.9%. In addition to this reduction, neural networks provide drivers with useful information about when the optimal moment is to charge the battery taking into account the renewable energy contribution.

b. Eco-design

This algorithm contributes to eco-design as:

b.1. It allows reducing emissions between 8.96 and 23.55 kg per day of CO_2 more than when EVs do not use the algorithm in this study.

b.2. This research shows that not only the number of EVs is important to increase the energy available but the way of choosing the engineers' locations. The algorithm proposed in this study allows establishing the optimal locations. Therefore, a design center could obtain more energy by using a specific number of EVs.

b.3. The contribution of V2G to building energy demand ranges between 0.5% and 1.3% when using the algorithm proposed in this study in a small design center.

c. V2G and V2H compatibility

Even though battery performance is not degraded due to charging and discharging processes, current policies keep the user from participating in V2G and V2H at the same time. The EV charging fee is higher than the savings obtained when using V2H. Consequently, V2G is not compatible with V2H. However, policies cannot be changed if the power of renewable energy installed is not increased. Therefore, renewable energy mix vs not renewable energy is not high enough. Consequently, the policies to promote EVs are as important as increasing the power of renewable energy installed.

d. Cybersecurity

As detailed in the cybersecurity section, some policies to assure that an electronic control unit (ECU) is not violated implies that the electronic control unit is no longer available. In this present study, 190 kg of waste is generated every year taking into account the electronic control units scrapped. Consequently, some techniques used for ECU cybersecurity are not eco-friendly, and more research should be done into this topic to better integrate cybersecurity and eco-design.

Appendix: How to Configure the Calls to Here® API

Here® API is able to provide a great deal of information about how to go from A to B. However, the calls to the Here® API must be done properly. In this research, the API is called by using the pseudo-code shown in Fig. 3.20. The string PARAM must be built. The following parameters are indicated:

A. apiKey. This key is generated when a user is registered in Here® developers' website. Replace XXXX by your key.

```
"""Routing the fastest"""
PARAM=("apiKey=XXXX" +
    "&waypoint0=geo!" + str(location_coor[1]) + "," + str(location_coor[2]) +
"&waypoint1=geo!"+
    str(location_coor[3]) + "," + str(location_coor[4]) +
"&mode=fastest;car;traffic:enabled&consumptionmodel=standard&" +
    "customconsumptiondetails=" +
    "speed,0,1.7,10,1.4,30,1.1,50,1.0,70,1.1,100,1.2,120,1.4,140,1.8;" +
    "ascent,30.0;descent,10.0;auxiliaryconsumption,0.8;acceleration,0.2;deceleration,0.3" +
   "&legAttributes=links,trafficTime&linkAttributes=consumption,dynamicSpeedInfo")
response[1]=routing(PARAM)

"""Routing the shortest"""
PARAM=("apiKey=XXXXXX" +
    "&waypoint0=geo!" + str(location_coor[1]) + "," + str(location_coor[2]) +
"&waypoint1=geo!"+
    str(location_coor[3]) + "," + str(location_coor[4]) + "&mode=fastest;car;traffic:enabled")
response[2]=routing(PARAM)
```

Fig. 3.20 Pseudocode to obtain the desired Here® answer

B. *Waypoint0* and *waypoint1* contains the latitude and longitude information about origin and destination locations which are stored in *location_coor* dictionary in Python. *Geopy* package can be used to obtain coordinates.
C. Mode. This parameter contains important information for Here® API such as the type of route (the fastest, the shortest) and traffic state.
D. A consumption model has to be indicated by using *consumptionmodel* and *customsonsumptiondetails.*

The response variable stores the answer from Here® server.

The answer from Here® is a json file. The reader can find a great deal of information such as speed estimation taking into account traffic conditions (*trafficSpeed* parameter), time elapsed for a specific speed and consumption, etc. The reader can easily estimate the average consumption in kWh, the average speed, etc. (Fig. 3.21).

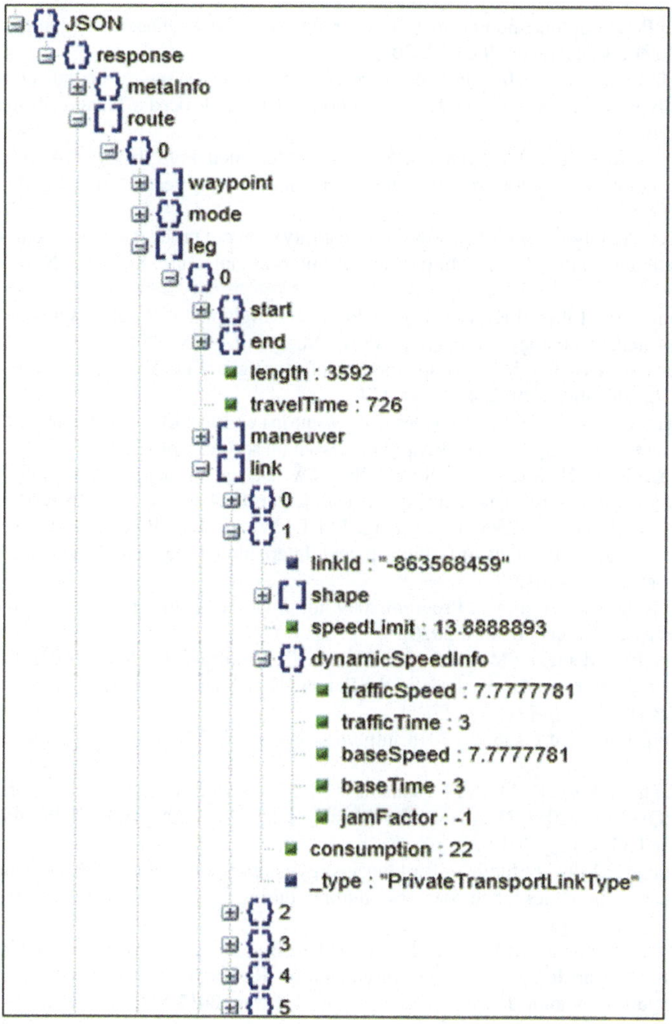

Fig. 3.21 Here® answer. Json format

References

1. European Union website https://ec.europa.eu/growth/industry/sustainability/ecodesign_en (accessed on second May 2020).
2. Sustainability Guide founded by European Union https://sustainabilityguide.eu/ecodesign/ (accessed on third May 2019).
3. Chun Y.; Lee K.; Lee J.S.; Lee J.Y.; Lee M.H.; Mishima N.; Tahara K. Identifying key components of products based on consumer- and producer-oriented eco-design indices considering environmental impacts, costs, and utility value. Journal of Cleaner Production **2018**, 198, 1031–1043.

4. Cicconi P. Eco-design and Eco-materials: An interactive and collaborative approach. Sustainable Materials and Technologies **2020**, 23, 135.
5. Götze U.; Peças P.; Richter F. Design for eco-efficiency – a system of indicators and their application to the case of moulds for injection moulding. Procedia Manufacturing **2019**, 33, 304–311.
6. Peiris R.L; Kulatunga A.K.; Jinadasa S.N. Conceptual model of Life Cycle Assessment based generic computer tool towards Eco-Design in manufacturing sector. Procedia Manufacturing **2019**, 33, 83–90.
7. de Grave, A.; Olsen. Challenging the sustainability of micro product development. 4M Second International Conference on Multi-Material Micro Manufacture, Springer: New York, USA, 2006.
8. Morgan, J.M.; Liker, J.K. The Toyota Product Development System: Integrating People, Process, and Technology; Productivity Press: New York, USA, 2006.
9. Rosen M.A.; Kishawy H.A.Sustainable Manufacturing and Design: Concepts, Practices and Needs. Sustainability **2012**, 4(2):154–174.
10. European Court of Auditors https://www.eca.europa.eu/Lists/ECADocuments/SR20_01/SR_Ecodesign_and_energy_labels_EN.pdf (accessed on second April 2020).
11. Seow, Y.; Goffin, N.; Rahimifard, S.; Woolley, E.A. 'Design for Energy Minimization' approach to reduce energy consumption during the manufacturing phase'. Energy **2016**, 109, 894–905.
12. Ka-Leung-Moon, K. ; Youn, C. ; Chang, J.M.T.; Yeung, A.W. Product design scenarios for energy saving: A case study of fashion apparel. International Journal of Production Economics **2013**, 146(2), 392–401.
13. United Nations Environment Program 2009 http://www.d4s-sbs.org/d4s_sbs_manual_site_S.pdf (accessed on second April 2020).
14. Tecchio, P.; Ardente, F.; Marwede, M.; Clemm, C.; Dimitrova, G.; Mathieux, F. Ecodesign of Personal Computers: An Analysis of the Potentials of Material Efficiency Options. Procedia CIRP **2018**, 69, 716–721.
15. Wolfram, Lago and Osborne, 2018 http://oro.open.ac.uk/52888/ (accessed on second April 2020).
16. Naumann, S., Kern, E., Dick, M. , Joham, T. 2015 https://files.ifi.uzh.ch/hilty/t/Literature_by_RQs/RQ%20120/2015_Naumann_Kern_Dick_Sustainable%20Software%20Engineering.pdf (accessed 3 February 2019).
17. Xiaodong C.; Xilei D.; Junjie L. Building energy-consumption status worldwide and the state-of-the-art technologies for zero-energy buildings during the past decade. Energy and Buildings **2016**, 128, 198–213.
18. Colin, C.V.; Capilla, R.; Betz, S.; Penzenstadler, B.; Crick, T.; Crouch, S.; Nakagawa, E.Y.; Becker, C.; Carrillo, C. Software sustainability: Research and practice from a software architecture viewpoint. Journal of Systems and Software **2017**,138, 174–188.
19. Nurcan, S.; Soffer, P.; Bajec, M.; Eder, J. Advanced Information Systems Engineering. 28th International Conference, CAiSE, Springer: New York, USA, 2016. pp. 35–36.
20. Ellis, G. Control System Design Guide, 1st ed., Springer: New York, USA, 2012.
21. Rajeev, R.; Kasun H.; Rehan S. Improving the energy efficiency of the existing building stock: A critical review of commercial and institutional buildings. Renewable and Sustainable Energy Reviews **2016**, 53, 1032–1045.
22. Bui, D.; Ngoc-Nguyen, T.; Ghazlan, A.; Ngo, N.; Duc-Ngo, T. Enhancing building energy efficiency by adaptive façade: A computational optimization approach. Applied Energy **2020**, 2020, 114797.
23. Zeyu, W.; Ravi, S.S. Classification of Household Appliance Operation Cycles: A Case-Study Approach . Energies **2015**, 8(9), 10522–10536.
24. Qi, X.; Wu, G., Barth, M.J.; Boriboonsomsin, K. Energy Impact of Connected Eco-driving on Electric Vehicles. In Road Vehicle Automation 4. 2nd Ed; Beiker, S.; Gereon, M.; Springer: Nueva York, United States, 2017; Volume 4, pp. 97–111.
25. HAL https://hal.archives-ouvertes.fr/hal-01405291/document (accessed on 12th March 2020).
26. http://www.d4s-sbs.org/d4s_sbs_manual_site_S.pdf (accessed on second April 2020).

27. Nunzio, G.; Thibault, L.; Sciarretta, A.; IEEE. International Conference on Intelligent Transportation Systems 2016. https://doi.org/10.1109/itsc.2016.7795927.
28. University of California. https://escholarship.org/uc/item/9z18z7xq October on 25th September 2019.
29. Haque, A.; Rahman, M.A. Study of a solar PV-powered mini-grid pumped hydroelectric storage & its comparison with battery storage. Conference: 7th Electrical & Computer Engineering (ICECE), 2012.
30. Haque, A.; Rahman, M.A.; Ahsan, Q. Building Integrated Photovoltaic system: Cost effectiveness. 7th International Conference on Electrical and Computer Engineering, 2012.
31. Tarroja, B.; Li Z.; Wifvat, V; Shaffer, B.; Samuelsen, S. Assessing the stationary energy storage equivalency of vehicle-to-grid charging battery electric vehicles. Energy **2016**, 106, 673–690.
32. Yang, Z.; Noori, M.; Tatari, O. Vehicle to Grid regulation services of electric delivery trucks: Economic and environmental benefit analysis. Applied Energy **2016**, 170, 161–175.
33. Uddin, K.; Dubarry, M.; Glick, M.B. The viability of vehicle-to-grid operations from a battery technology and policy perspective. Energy Policy 2018. 113, 342–347.
34. Sovacool, B.; Noel, L.; Axsen, J.; Kempton, The neglected social dimensions to vehicle-to-grid (V2G) transition: a critical and systematic review. Environmental Research Letters. https://iopscience.iop.org/article/10.1088/1748-9326/aa9c6d
35. HERE. https://developer.here.com/ accessed on 29th March 2020.
36. HERE. https://developer.here.com/documentation/routing/dev_guide/topics/resource-param-type-custom-consumption-details.html accessed on 29th March 2020.
37. De Cauwer C.; Van Mierlo J.; Coosemnans T. Energy Consumption Prediction for Electric Vehicles Based on Real-World Data. Energies **2015**, 8, 8573–8593.
38. Palma-Mendez, J.T.; Marín-Morales, R. Inteligencia Artificial, 2nd edition.; Mc Graw Hill: New York, USA, 2008; pp. 647–691.
39. Hagan, M.T.; Demuth, H.B.; Beale, M.H.;de Jesús, O. Neural Network Design, 2nd ed.: Martin Hagan editors: Oklahoma. USA. 2012; 14.1–14.48.
40. Mathworks https://es.mathworks.com/help/deeplearning/ref/narnet.html;jsessionid=1b35de9b0f7625f9bcc764bb5ac5 Accessed on 28th January 2020.
41. Ponce-Cruz, P. Inteligencia Artificial, 1st ed.: Marcombo: Barcelona, Spain. 2011; pp. 238–265.
42. Russel, E., Yuhui, S. Computational intelligence, 2nd ed: Morgan Kaufmann : San Francisco, USA, 2011: pp. 197–265.
43. French System Operator https://www.rte-france.com/fr/eco2mix/eco2mix-co2 Accessed on 15th February 2020.
44. Baca-Ruiz, L.G.; Pegalajar-Cuéllar, M.; Delgado Calvo-Flores M.; Pegalajar-Jiménez M. An Application of Non-Linear Autoregressive Neural Networks to Predict Energy Consumption in Public Buildings. Energies **2016**, 9, 684.
45. French Ministry of Ecologic transition https://www.statistiques.developpement-durable.gouv.fr/sites/default/files/2019-05/datalab-53-chiffres-cles-des-energies-renouvelables-edition-2019-mai2019.pdf Accessed on 8th May 2020.
46. The International Council on Clean Transportation. https://www.sciencedirect.com/ Accessed on 8th May 2020.
47. Kester, J.; Noel L.; Zarazua de Rubens G.; Sovacool B.K. Promoting Vehicle to Grid (V2G) in the Nordic region: Expert advice on policy mechanisms for accelerated diffusion. Energy Policy 2018, 116, 422–432.

Chapter 4
Driver Efficiency and Software: Influence on Vehicle-to-Building

David Borge-Diez, Pedro-Miguel Ortega-Cabezas, Antonio Colmenar-Santos, and Jorge-Juan Blanes-Peiró

Abbreviations

API	Application Programming Interface
EC	Eco-charging
EDR	Eco-driving
ER	Eco-routing
GRU	Gated Recurrent Units
NAR	Nonlinear Autoregressive Neural Networks
RE	Renewable Energy
V2B	Vehicle-to-Building
V2G	Vehicle-to-Grid
V2H	Vehicle-to-Home
VCU	Vehicle Control Unit

Extract from *Energies* **2021**, *14*(12), 3483; https://doi.org/10.3390/en14123483, https://www.mdpi.com/journal/energies.

D. Borge-Diez · J.-J. Blanes-Peiró
Department of Electrical and Control Engineering, Universidad de León, Campus de Vegazana, s/n, 24071 León, Spain
e-mail: david.borge@unileon.es; dbord@unileon.es

J.-J. Blanes-Peiró
e-mail: jorge.blanes@unileon.es

P.-M. Ortega-Cabezas · A. Colmenar-Santos (✉)
Departamento de Ingeniería Eléctrica, Electrónica y de Control, UNED, Juan del Rosal, 12, Ciudad Universitaria, 28040 Madrid, Spain
e-mail: acolmenar@ieec.uned.es

P.-M. Ortega-Cabezas
e-mail: pedro-miguel.ortega-cabezas@valeo.com

© The Author(s), under exclusive license to Springer Nature Switzerland AG 2024

A. Colmenar-Santos et al. (eds.), *Development and Testing of Vehicle Software and its Influence on Sustainable Transport*, https://doi.org/10.1007/978-3-031-47630-3_4

4.1 Introduction

Energy consumption in the transport sector and in buildings are of great concern. This research aims to quantify how eco-routing (ER), eco-driving (EDR) and eco-charging (EC) can increase the amount of energy available for vehicle-to-building (V2B). To do this, the working population was broken into social groups (freelancers, local workers and commuters) who reside in two cities with different climate zones (Alcalá de Henares-Spain and Jaén-Spain) as the way of using electric vehicle (EV) is different. An algorithm based on the Here® application program interface and neural networks was implemented to acquire data of the stochastic usage of EVs of each social group. Finally, an increase in the amount of energy available for V2B thanks to the algorithm was assessed.

The results per day were as follows. Owing to the algorithm proposed a reduction ranging from 0.6 kWh to 2.2 kWh was obtained depending on social groups. The proposed algorithm facilitated an increase in energy available for V2B ranging from 13.2 kWh to 33.6 kWh depending on social groups. The results display that current charging policies are not compatible with all social groups and do not consider the renewable energy contribution to the total electricity demand.

When it comes to Spain, the transport sector is the main culprit for discharging pollutants into the atmosphere as well as consuming vast amounts of energy [1]. Recently, the private research center named Economics for Energy has confirmed this statement in its latest transport report as this trend will certainly continue in the year to come due to mobility needs [1]. As published by the Spanish Government, the energy consumption linked to transport has been increasing since 2013, just after the economic crisis from 1,074,714 TJ to 1,196,381 TJ in 2018 (latest data available) [2]. The same trend can be found for CO_2 emissions which accounted for 115,402,074 t in 2013 and 128,275,075 t in 2020 [2].

Electric mobility plays a key role when reducing greenhouse emissions as shown by Pillay et al. [3]. In their research, they showed how emissions could be reduced in South Africa up to 12.3% considering a specific e-car, e-truck and e-bus penetration in the market. Bastida-Molina, Hurtado-Pérez and Peñalvo-López drew similar conclusions proving that emissions can be reduced by increasing the number of EVs and boosting the number of MW of renewable energy (RE) [4]. As they stated in their research, a 100% RE generation will be needed in order to reduce up to 74 million tons per year. Zheng et al. display in their research that, *from 2011 to 2017, 682,047 plug-in EVs were sold in 5 provinces of China, with 18.3 billion electric vehicle kilometers traveled, 3.0 TWh of electricity consumed, a reduction in gasoline consumption of 1.6 billion liters and in CO_2 emissions by 611,824 tons* [5]. Despite these results, some research is focused on the impact of EVs on the environment and tries to evaluate whether they are more sustainable than traditional powertrains such as internal combustion engine vehicles. Helmers et al. made an interesting study comparing the environmental impacts of petrol, diesel engines and natural gas engines as well as EVs. They proved that producing battery cells with renewable electricity decreases the environmental impacts of EVs considerably [6]. In addition, EV impacts can be reduced even more by making a better use of mineral resources [6]. Messagie et al. proved that EVs are the most sustainable means of transport taking into account the whole life cycle [7].

EVs are useful not only for being zero-pipeline vehicles but for being a key element in several important techniques such as vehicle-to-grid (V2G), vehicle-to-home (V2H) and, generally speaking, vehicle-to-X connections [8–11]. V2G aims to use EVs as virtual power plants in such a way that the energy stored in EV batteries could be injected into the electricity network when needed in order to reduce consumption peaks and emissions among others [12]. As detailed by Bibak and Tekiner-Mogulkoç, today V2G is facing several barriers and obstacles such as high investments needed to apply this technology, stochastic nature of EVs (arrival and departure times, km covered, etc.), social issues and, finally, battery degradation [13]. V2G success is based on the participation of the EV owner. Consequently, the willingness to pay is an essential concept that has already been discussed in some research [14, 15]. This willingness can be defined by the maximum price that can stimulate an EV owner to inject the energy available of the EV battery into the grid. Other approaches linked to social issues were introduced by Noel et al. In their dissertation, they proved how concepts such as tinkering, testing and tacit knowledge[1] may accelerate the adoption of V2G [16]. Therefore, policies are an essential topic as discussed in several research [17, 18]. An important advantage of V2G is its capacity of better integration of the RE into the system as detailed by Mwasilu et al. [19]. At this point, it is vital to consider that the usage of V2G must be reliable, in other words, it must guarantee the system reliability, and EVs can play an important role to ensure it. To assure this, research has proposed the usage of artificial intelligence. Rahbari et al. proposed a solution based on a neuro-fuzzy inference system in order to integrate better REs and EVs into the grid considering generation source intermittency and energy usage inconsistency [20]. Mozafar, Moradi and Amini proposed a genetic algorithm-particle swarm optimization algorithm aiming at reducing power losses, voltage fluctuations, charging and demand supplying costs as well as EV battery costs [21]. Finally, it must be considered that battery degradation is due to V2G participation. Recent studies show that calendar ageing is influenced by factors such as standing time, state of charge and temperature whereas cycling ageing is affected by cycle number, depth of discharge and charging rate [22].

V2B is an important topic analyzed in a lot of research. It aims to reduce emissions and energy consumption of buildings owing to the energy stored in the batteries of EVs. An important reduction of fossil electricity is obtained due to the RE contribution when charging EV batteries whose energy will be transferred to the building as detailed by Buonomano et al. [23]. Zhou et al. describe in their research the main advantages of integrating EVs with RE such as cutting energy consumption of buildings, reducing the import/export pressure on the electric grid and shifting peak-loads

[1] As discussed by Noel et al., tinkering is defined as "*user modification of a technology to develop new innovations and uses*". Testing "*deals intently with experimentation in technology, especially between the designers of a product or artifact and its users*". Finally, tacit knowledge "*is the embodied knowledge a user may have in learning to utilize or modify a technology*".

to sub-peak or off-peak periods [24]. The main technical issue that this technology faces is the stochastic characteristic of driving schedule of EVs [25, 26].

All aforementioned techniques are based on EVs and, if their energy efficiency was improved, the amount of energy available to be used would be increased. ER and EDR are key elements to improve energy efficiency. ER consists of finding the most energy-efficient route for a vehicle to travel between two points in such a way that an optimal way to reduce energy consumption is offered to drivers. Thibault and Sciaretta proposed an energy consumption model which considers speed fluctuations and road infrastructure to reduce consumption [27]. Some elements such slopes have a significant influence on energy consumption and are considered in some ER algorithms [28]. Other authors have proposed the usage of evolutionary algorithms [29]. When it comes to EDR and pollutants, research is focused on this topic no matter what types of powertrain are used. Orfila, Saint-Pierre and Messias proposed an android application based on EDR assistance for internal combustion engines [30]. Similar research can be found for hybrid and EVs [31, 32].

4.2 Methods

4.2.1 Description

The main idea behind this method is to choose two representative cities, buildings and residents of those buildings in order to assess the improvements introduced by EDR, ER and EC based on an algorithm implemented on Python and neural networks (Fig. 4.1). Firstly, the cities subjected to this study are chosen considering climatic zones, traffic conditions and population among others. Secondly, the buildings used to measure improvements in energy consumption when using EDR and ER are selected. The participants, who are residents in the chosen buildings, are determined. The energy consumption estimates of the buildings are made, and, owing to the algorithm, the improvements are assessed [33].

4.2.2 Choice of the City

Spain has different climatic zones but, as shown in Fig. 4.2, two of them prevail in the country: the Mediterranean and Continental ones. Consequently, the consumption patterns are different. The former is characterized by hot dry summers and mild winters. This climate can be broken down into three subtypes. The latter is characterized by wide diurnal and seasonal variations in temperature and by low, irregular rainfall with high rates of evaporation that make the land arid. To do this study more representative, two cities belonging to different climatic zones were chosen: Alcalá de Henares and Jaén (Figs. 4.3 and 4.4).

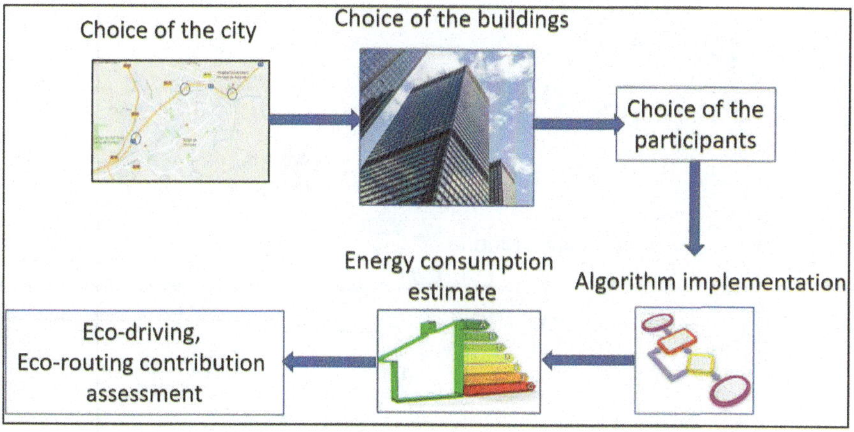

Fig. 4.1 Methodology used in this research

Fig. 4.2 Spanish climate

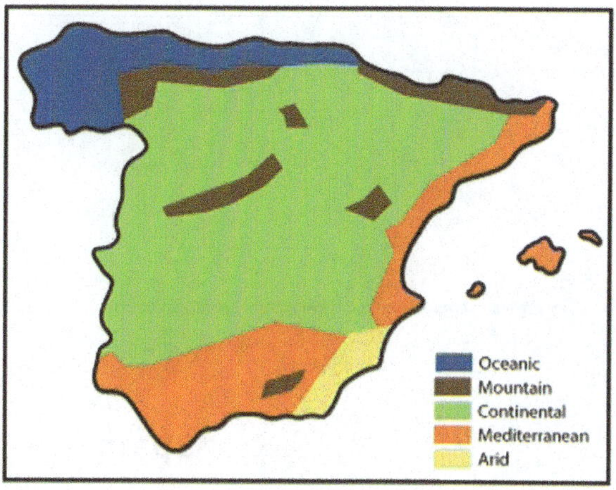

Alcalá de Henares belongs to the Community of Madrid (Spain), 32 km away from the Spanish capital (Fig. 4.3). To be more specific, this city is the second biggest one in this Community. Its population is 197,345 inhabitants in 2020 based on the data provided by the Spanish National Institute of Statistics. Generally, traffic jams are often reported in the city. Aiming at monitoring the traffic state, the townhall has installed 17 cameras mainly located downtown. This city is also interesting as it is connected to one of the most important roads to enter Madrid.

The population of Jaén which is located in the south of Spain is 112,757 in 2020 according to the data provided by the Spanish National Institute of Statistics (Fig. 4.4). This city has installed cameras in strategic points aiming to improve the traffic in the city due to its high intensity.

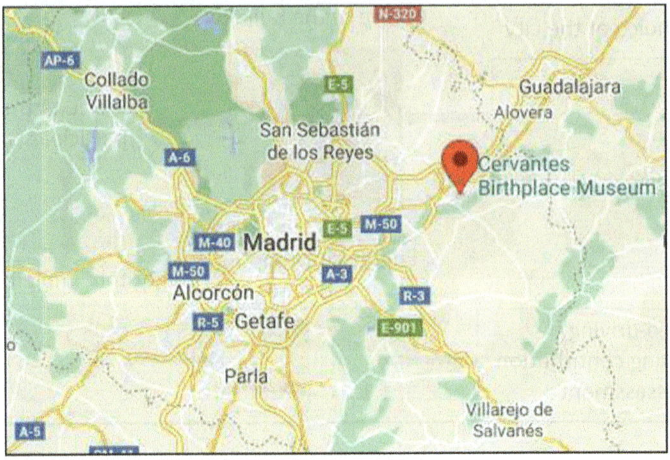

Fig. 4.3 Alcalá de Henares location

Fig. 4.4 Jaén location

Taking into account all the data mentioned above, a comparison between two similar cities located in different climatic zones can be made.

4.2.3 Choice of the Buildings

Several factors influence energy consumption when it comes to buildings. As detailed by Huebner et al., different variables such as building factors, socio-demographics, attitudes and self-reported behavior in their research affect energy consumption in different buildings [34]. To be more specific, 39% of the variability in energy consumption is linked to buildings, 24% to social-demographic variables, 14% is linked to heating behavior and only 5% is linked to attitudes and other behaviors [34]. This study was based on a sample of 924 English households.

Hueber et al. did also research about the aforementioned topic focused on electricity consumption instead of energy consumption by using an 845 English household sample [35]. In their research, they showed that 34% of variability in electric consumption came from appliance use and lighting while 21% was linked to social-demographic variables. Harputlugil and de Wilde analyzed in-depth the interaction between buildings and humans when it comes to energy consumption. In this case, one important conclusion is that lifestyle is an essential factor to be considered to understand energy consumption profiles and occupants' patterns [36]. Pan et al. concluded the importance of having a good understanding of how occupants use the appliances of buildings in order to improve energy efficiency [37]. Moreover, like this the occupants' behavior could be addressed to a more sustainable one. Finally, Yousefi, Gholipour et Yan proved that occupants' behavior and envelope materials play a key role when choosing the envelop material types [38].

Considering this, the conclusions that can be drawn when choosing buildings for this study are:

a. Orientation of buildings is a key factor when it comes to electricity consumption and energy in general.
b. Based on the previous studies, a different occupation rate for each flat must be considered.
c. As EVs will be used to assess the energy available for V2B technology, different social sectors must be considered.
d. Finally, different work timetables should be considered in this study.

4.2.4 Choice of the Participants in This Research

In this study, the authors have proceeded to choose participants according to their professions and the number of kilometers covered. This is a key point to be taken into account in order to obtain more accurate results. Based on this, one can distinguish:

- Freelancers. They cover a considerable number of kilometers on a daily basis according to statistical data published by different institutions [39]. Consequently, they are supposed not to inject a large amount of energy when using the V2B technology.
- Commuters. Commuters can choose between using the public transport and their own vehicles to get to work. In this study, commuters are supposed to drive to work by using EVs. Thanks to statistical methods described earlier, the daily mileage and energy consumption will be assessed.
- Local workers. Again, local workers can use the public transport or their vehicles to get to work. In this study, local workers who use their vehicles to go to work have been considered. In this case, the number of kilometers covered is supposed to be low. Consequently, they can contribute in a very important way when using the V2B technology.

All assumptions made about kilometers covered on a daily basis have to be confirmed based on statistical methods. Consequently, the reduction of emissions and energy consumption is influenced by the type of workers mix living in a building. The number of participants is shown in Table 4.1: freelancers, local workers, commuters and other occupants of the household who do not belong to these social groups (mainly students and the unemployed). Potential deviations in the assumptions earlier described are considered in the sensitivity analysis.

The algorithm was described in the previous chapter.

4.2.5 Statistical Analysis

To establish the average consumption of EVs for each social group, a statistical analysis must be done. Owing to this, the energy available for V2B is assessed for freelancers, local workers and commuters. As per the first analysis of the data collected, these data were closed to a normal distribution. The R software, and more specifically the PASSWR package, was used to confirm this assumption [40]. The main advantage of this package is that the data can be explored in depth thanks to the statistical parameters such as kurtosis, skewness and p-value. Kurtosis is a measure of relative peakedness of distribution. It is a shape parameter that characterises the degree of the peakedness. A distribution is said to be leptokurtic when the degree of peakedness is higher than 3; it is mesokurtic when the degree of peakedness is equal to 3, and it is platykurtic when the degree of peakedness is less than 3 [41, 42]. Skewness refers to a distortion or asymmetry that deviates from the symmetrical bell curve or normal distribution in a set of data. When the data under analysis are close to a normal distribution, the skewness is close to 0. Symmetry, it is vital to note, does not infer that the data follow a normal distribution. Consequently, the aforementioned analysis must be carefully conducted. The p-value is the probability of finding the observed, or more extreme, results when the null hypothesis (H_0) of a study question is true.

Table 4.1 Number of participants in this research

Building	Location	Orientation	Average surface (m^2)	Occupants	Total participants	Commuters	Local workers	Freelancers	Rest of occupants
A	Alcalá	South	90	35	15	5	2	8	20
B	Alcalá	North	85	45	20	8	3	9	25
C	Alcalá	West	100	32	15	6	2	7	17
D	Jaén	South	75	42	20	10	2	8	22
E	Jaén	North	80	45	20	11	2	7	25
F	Jaén	West	90	12	10	5	1	4	2

In addition to the statistical parameters described earlier, several plots were used to confirm that the data analysed follow a normal distribution: histograms, Q-Q plots and boxplot. The histogram represents the frequency of occurrence of specific phenomena that lie within a specific range of values, which are arranged in consecutive and fixed intervals. The quantile–quantile or q-q plot is an exploratory graphical device used to check the validity of a distributional assumption for a data set. A boxplot, sometimes called a box and whisker plot, is a type of graph used to display patterns of quantitative data [41].

4.3 Results

4.3.1 EV Consumption

The algorithm proposed in this research aims to improve energy efficiency by considering the stochastic usage of EVs. To do this, many measurements were performed on social groups such as freelancers, local workers and commuters. To assess each social group's average energy consumption, the collected data were statistically analysed. The results obtained are shown in Table 4.2 for Alcalá de Henares. First of all, it must be taken into account that all these data are close to a normal distribution. This assumption was confirmed by assessing different parameters such as skewness, kurtosis and *p*-value. When it comes to freelancers, skewness is close to zero. Consequently, the distribution is symmetric. Kurtosis aims to prove that the data distribution tails are not dissimilar from normal distribution ones. The *p*-value represents the null hypothesis: the data follow a normal distribution. When the *p*-value is greater than 0.05, the null hypothesis is confirmed. One important characteristic of skewness and kurtosis is that they are sensitive to the sample size. Therefore, the Q-Q plots and histograms were used to confirm that the data collected followed a normal distribution. Based on this statistical analysis, the hypothesis H_0 was confirmed. In addition to energy consumption, it is essential to establish the number of kilometres by following the same statistical method. Regarding freelancers, the number of kilometres was found to range from 95 to 110 km per day. When it comes to local workers and commuters, the number of kilometers covered varies from 57 to 70 km per day. Finally, commuters cover between 4 and 6 km per day. When it comes to Jaén, the number of kilometers travelled by freelancers ranges between 78 and 90 km a day (Table 4.3). Regarding commuters, they drive between 3.5 km and 5.5 km daily. Finally, local workers cover between 3.2 km and 5.3 km a day. The main differences between both cities are analysed in Sect. 4.4.1. These results are complementary to other research. Zhang et al. detailed how large-scale EV development impacts the stability of electric grid as well as decisions linked to the construction of new facilities (charging facilities). They conducted a depth-study of stochastic usage of EVs based on several factors such as daily distance travelled, energy consumption, etc. [43]. Similar research was done by Shi et al. showing interesting data about the

Table 4.2 EV consumption in kWh in Alcalá de Henares

Factor	Freelancers		Commuters		Local workers	
	A.U.[a]	N.A.[b]	A.U.[a]	N.A.[b]	A.U.[a]	N.A.[b]
Mean	24	26.2	9	10.5	3.3	3.9
Std deviation	0.6	0.4	0.3	0.32	0.32	0.29
Kurtosis	3.7	4.0	3.7	4.5	4.1	4.2
Skewness	−0.135	−0.121	−0.041	−0.032	−0.025	−0.015
p-value	0.395	0.401	0.401	0.415	0.396	0.399

[a]N.A. means no algorithm is used
[b]A.U. means the algorithm is used

Table 4.3 EV consumption in kWh in Jaén

Factor	Freelancers		Commuters		Local workers	
	A.U.[a]	N.A.[b]	A.U.[a]	N.A.[b]	A.U.[a]	N.A.[b]
Mean	21	22.9	8.5	9.5	4.3	4.9
Std deviation	0.7	0.3	0.4	0.36	0.36	0.31
Kurtosis	3.6	4.1	3.5	4.3	4.0	4.1
Skewness	−0.145	−0.111	-0.045	−0.03	−0.02	−0.013
p-value	0.385	0.301	0.411	0.405	0.356	0.349

[a]N.A. means no algorithm is used
[b]A.U. means the algorithm is used

average speed and consumption based on stochastic usage of EVs [43]. The impact on the electricity grid can also be stochastic as shown by Schey et al. [44]. All this research did not consider social groups which are a key element for future energy policies due to the fact that their usage of EVs is different as proved in this study.

4.3.2 Energy Consumption of Buildings

In order to verify the improvements introduced by the algorithm based on EC, ER and EDR, it is essential to measure energy consumption of the buildings chosen for this case-study. To do this, smart counters and power meters were installed in each participant's apartment. Consequently, it was possible to measure the energy consumption of all facilities of the apartments such as: electric heating, air conditioning, lighting, TV sets, fridges, microwave ovens, vitroceramic hobs, washing machines and dishwashers. All these measurements were performed in the climatic areas described in this chapter.

When analyzing the Spanish electricity demand curve, there are two main consumption peaks. The first one is usually between 12 a.m. and 1 p.m. and the other one from 8 p.m. to 10 p.m. depending on the season. There are two main differences between the summer and the winter curves. Firstly, the electricity demand is lower in the summer. Secondly, the peak between 12 a.m. and 1 p.m. is higher than the other one in the summer. Therefore, it is important to check the energy consumption pattern of the buildings chosen in this case study. Figure 4.5 shows the winter energy consumption for buildings A, B and C which are located in Alcalá de Henares. As one can see, the energy consumption trend is similar to the Spanish one as there are two peaks present. However, there are important differences. Firstly, from 4 p.m. to 6 p.m. the energy consumption decreases as the energy consumption patterns of buildings A and C follow the Spanish one. Nevertheless, it is not the case for building B. Secondly, the trend followed by buildings A and C has similarities comparing to the Spanish trend, but this first energy consumption peak lasts less time than expected. Figure 4.6 shows results during summer time. As expected, two consumption peaks are present and the first one is the most important. Again, in this case, building B has a different behavior in comparison with buildings A and C as its consumption is higher during the off-peak periods. As explained in the discussion section, these particularities of electricity demand curves can be analyzed considering the number of occupants of the apartments as well as the diversity of social groups they belong to. The reader can find the curves for the second climatic zone in the supplementary data as the conclusions are similar except for the fact that total energy consumption changes slightly. This variation is totally normal as in Jaén the consumption is slightly lower as stated by the Institute of Diversification and Energy Saving of Spain.

Variations of energy consumption depending on climatic zones have been studied widely in the scientific literature [45–48]. Energy consumption profile is an essential topic to be considered as shown in a lot of research. Csoknyai et al. analyzed in detail energy consumption and household composition (couple with children, couples without children, single, single with children and other) [49]. Kavousian,

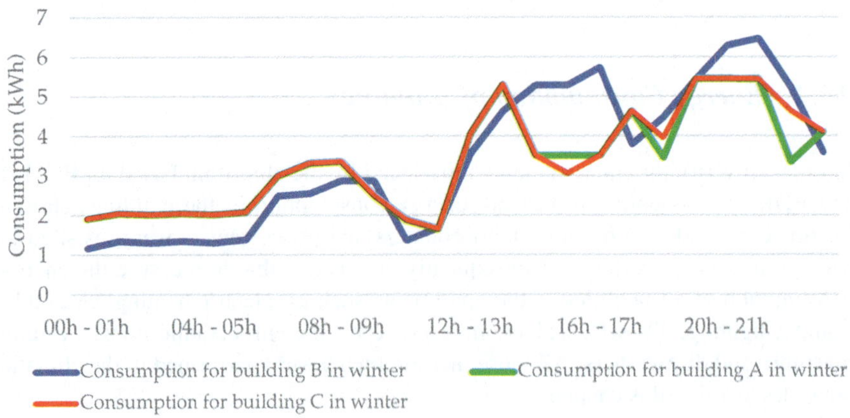

Fig. 4.5 Winter consumption curves for the buildings located in Alcalá

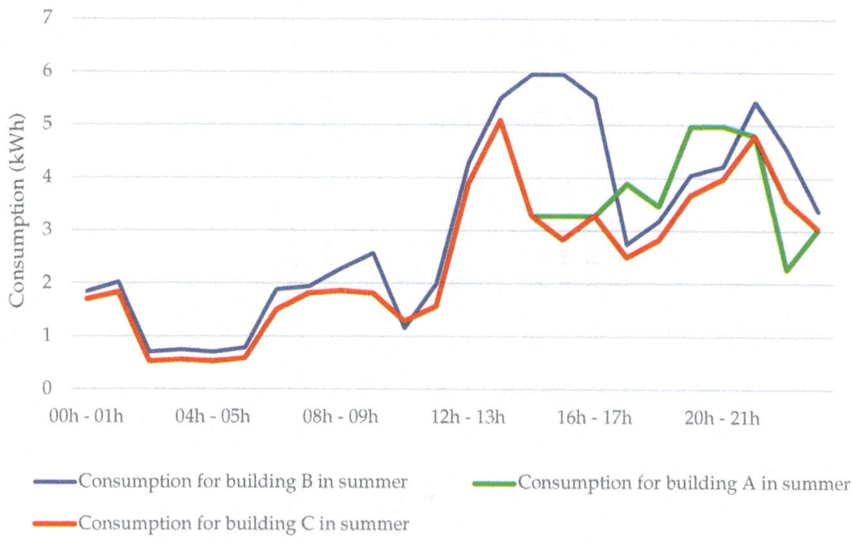

Fig. 4.6 Summer consumption curves for the buildings located in Alcalá

Rajagopal and Fisher focused their efforts on analyzing pattern energy consumptions of buildings based on classifying the occupants into age groups [50]. Laaroussi et al. conducted research into how occupant presence and behavior influence energy consumption [51]. However, the research does not take into account social groups, occupants present, EDR, EC and ER at the same time. Table 4.4 quantifies how energy consumption of the building can be reduced when all these factors are considered at the same time. As one can see, the contribution to V2B is different for each building due to the number of freelancers, local workers and commuters. Consequently, it is essential to discuss this topic the way it is done in the next section. Although the results depicted in these two tables imply that all participants in this research contribute to V2B, this an assumption that might be false.

4.4 Discussion

4.4.1 EV Consumption

The energy consumption of EVs is stochastic as already proved in professional literature [43, 44]. This research aims to show that social groups are so important that they should be considered when analysing the stochastic usage of EVs. Freelancers covered many kilometres on a daily basis. Firstly, the results for Alcalá de Henares show that even though freelancers' contribution to V2G and V2B may not be important, ER and EDR can improve energy efficiency up to 8.4%. Due to the

Table 4.4 Energy available for each building in Alcalá de Henares and Jaén on a daily basis

	Freelancers			Local workers			Commuters			Energy available for V2B A.U.	Energy available for V2B N.A.[b]
	Number	kWh available A.U.[a]	kWh available N.A.[b]	Number	kWh available A.U.[a]	kWh available N.A.[b]	Number	kWh available A.U.[a]	kWh available N.A.[b]		
Building A	8	128	110.4	2	73.4	72.2	5	155	147.5	356.4	330.1
Building B	9	144	124.2	3	110.1	108.3	8	248	236	502.1	468.5
Building C	7	112	96.6	2	73.4	72.2	6	186	177	371.4	345.8
Building D	8	152	136.8	2	71.4	70.2	10	315	305	538.4	512
Building E	7	133	119.7	2	71.4	70.2	11	346.5	335.5	550.9	525.4
Building F	4	76	68.4	1	35.7	35.1	5	157.5	152.5	269.2	256

[a]N.A. means no algorithm is used
[b]A.U. means the algorithm is used

fact that freelancers are a much consuming energy social group when it comes to EV usage, they are forced to participate in V2G or V2B but not in both of them. Regarding commuters, the energy efficiency enhancement can reach 14.3% owing to the algorithm proposed in this study. In addition, the amount of energy available for V2B participation is high and owing to ER and EDR it is even higher. Finally, local workers would be the biggest contributors to V2B if they had an intention to buy EVs. However, EVs high prices can pose a serious obstacle for a large number of local workers. It must also be stated that the energy efficiency gain for this social sector is low (0.6 kWh). These data are valuable for policy makers as they show the social groups which might contribute most to V2G and V2B technologies. When it comes to Jaén and using the algorithm, one can find improvements in energy efficiency up to 8.1%, 10.5% and 12.2% for freelancers, commuters and local workers respectively. It is important to remark that the number of kilometres covered in both cities for each social sector are similar. After having analysed the data collected during the trips, two conclusions are drawn. Firstly, people are more likely to get caught in traffic jams in Alcalá de Henares than in Jaén. Consequently, a higher consumption is expected. Secondly, the difference between both cities are more remarkable in the winter. As Alcalá is colder than Jaén, EV performance is affected by temperature. This conclusion is aligned with other research. Sagaria, Neto and Baptista proved that energy consumption can vary between 25 and 30% depending on the location where the EV is used- in the northern or southern countries [51]. On the other hand, the difference between both cities is not big enough in order to find significant distinctions in EV charging patterns contrary to the conclusions drawn by Yan et al. [52].

4.4.2 Energy Available for V2B

The number of kWh available for V2B purposes depends on three factors. The first one is the share of EVs in the market. The second one is energy efficiency linked to EVs. This parameter should be improved by means of EDR and ER in order to increase the amount of energy available to be used for V2B. Finally, the third one is the social groups to which the participants belong to, as the way of using EVs is completely different. This research is focused on the two last concepts. As shown in Fig. 4.7, in the buildings chosen for this case-study, most of the people belong to the social groups of freelancers and commuters. Buildings D and E are the ones which achieve more significant savings in energy due to the enhancement in terms of EV energy consumption discussed in the previous section. As detailed earlier, freelancers fail to contribute in a significant way owing to the number of kilometers covered. However, even if local workers are the less important group, in some occasions such as the case of building B, they can contribute in a very important way as the number of kilometers covered is very low and, consequently, the energy available for V2B is high. When it comes to the usage of the algorithm proposed in this research, improvements which range from 13.2 kWh to 33.6 kWh can be obtained on a daily basis based on the results depicted in Fig. 4.8.

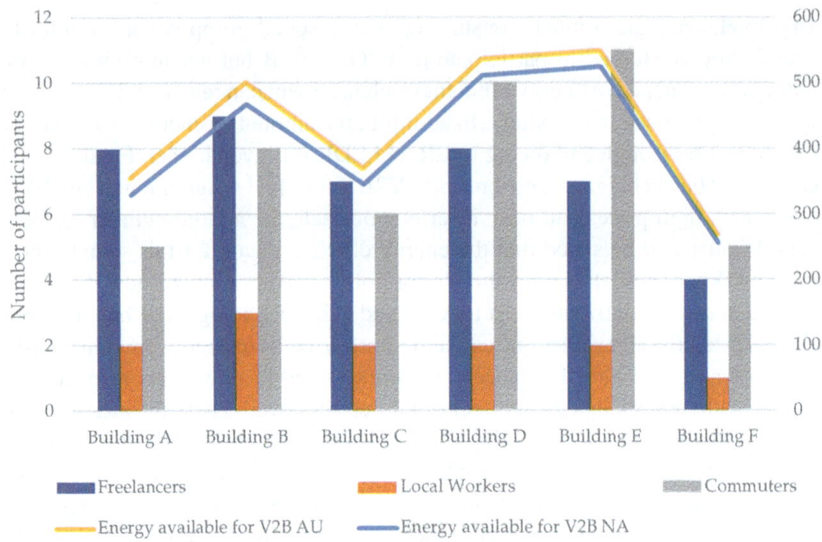

Fig. 4.7 Contribution to V2B based on social groups

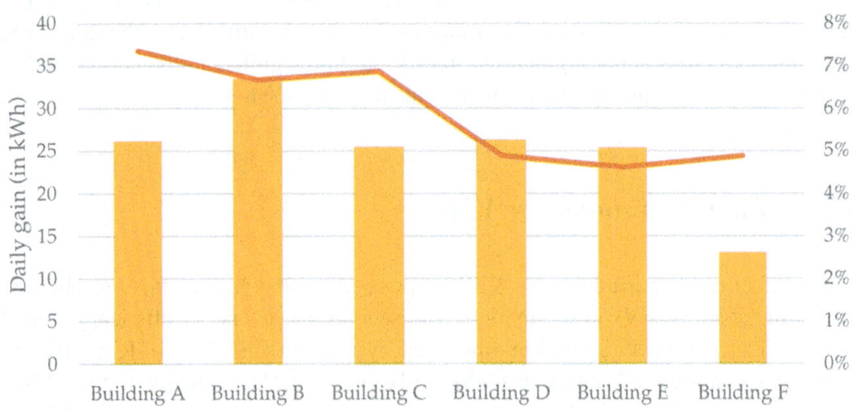

Fig. 4.8 V2B improvement owing to ER and EDR

Considering Figs. 4.7 and 4.8, commuters and local workers are the most important contributors to V2B. Consequently, policies should be addressed in order to increase the EV presence in these two social groups. Of course, freelancers are an important group to be considered based on an environmental point of view to reduce emissions but not based on their potential contribution to V2B.

The energy consumption which took place when freelancers, commuters and local workers were outside was not considered. Therefore, the household composition studied by Csoknyai et al. is not considered [49]. As shown in Fig. 4.9, local workers make the first energy consumption peak last more time (Buildings A and B). Secondly,

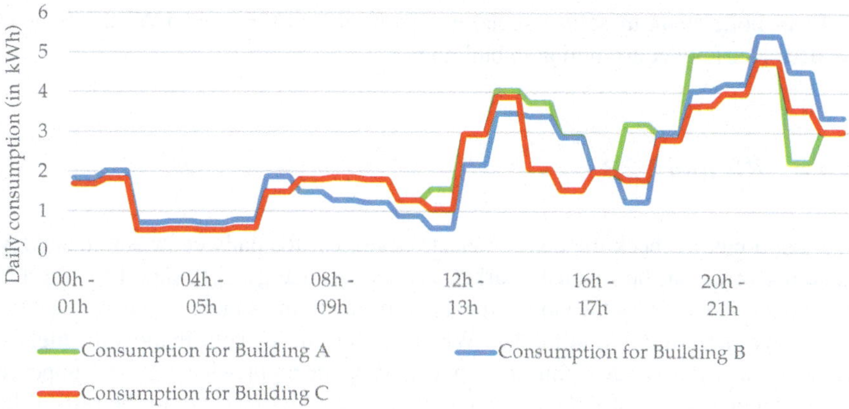

Fig. 4.9 Electric energy consumption considering social groups

local workers and commuters make the second consumption peak earlier. All these consumption peaks can be reduced due to EDR and ER algorithm.

4.4.3 *Social Group Presence in Buildings*

During this research, the authors have established which social groups contribute more to V2B. Consequently, it is of paramount importance to establish the percentage of people who belong to each social group. A binomial distribution was employed to determine the percentage of workers who belong to each social group:

a. Set an initial hypothesis based on the number of people belonging to each social groups. To do this, a sample of 5 buildings for each city was chosen.
b. A second sample was used in order to confirm or reject the hypothesis by using Eqs. (4.1) and (4.2):

$$H_0 \text{ is true if } \quad \frac{|\hat{p} - p_0|}{\sqrt{\frac{p_0(1-p_0)}{n}}} \leq Z_{\frac{\alpha}{2}} \tag{4.1}$$

$$H_0 \text{ is false if } \quad \frac{|\hat{p} - p_0|}{\sqrt{\frac{p_0(1-p_0)}{n}}} > Z_{\frac{\alpha}{2}} \tag{4.2}$$

n is the sample size, $\hat{p}$ is the probability of success for the sample considered, p_0 is the probability of confirmation of the hypothesis, and α is the significant level.

The results obtained are shown in Tables 4.5 and 4.6. Based on the data obtained in these tables and the consumption estimate shown in Tables 4.2 and 4.3, it is important to focus policies on these two social groups in order to increase kWh available to reduce electricity consumption in buildings.

4.4.4 RE and V2B

It is important to check the trend of electricity price, RE daily contribution and EV charging price on the Spanish market. As shown in Figs. 4.10 and 4.11, the EV charging price is set by the Government and from 1 p.m. is more expensive than the electricity price in the sport market. When it comes to RE contribution, the highest value is reached between 1 p.m. and 8 p.m. During the night, when EVs are supposed to be charged, the contribution is not extremely high comparing to the rest of the day. EVs are charged when RE contribution and the price are low. This policy is contradictory as EVs contribute to the reduction of pollution because their energy may be used to reduce the peak energy consumption from 7 p.m. to 10 p.m. but to do this, EVs must be charged before 12 a.m. At that moment, the RE contribution is the lowest. The main conclusion is that promotion of RE to decarbonize the electricity system, and policy prices are as important as policies to promote EVs penetration into the market or charging points implementation. Like this, the charging process can be performed when RE contribution is high. Regarding RE in Spain, the number of MW available has been almost stable since 2012. It must be stated that establishing the optimal sizing of RE facilities under high EV integration is a relevant topic studied in several research [53].

One important point is that V2B must be compatible with other techniques such as V2G. In this case, some important remarks must be made. Based on the total amount of energy available to be used for V2G, freelancers would tend to participate in V2B rather than in V2G, as they get more profit when using this energy to reduce their home energy consumption. When it comes to local workers, they could participate in V2G and V2B. The main issue is that they are not the most representative social sector. Finally, commuters seem to be a social sector that could also take advantage of both technologies. In order to extend the number of people who could participate in V2G and V2B, the fee to charge EVs in the second off-peak consumption should be reduced similar to Fig. 4.12. To do this, RE should be promoted and increased in order to support energy demand to charge EVs. Some research deals with the RE integration topics when using EVs. Pearre and Swan concludes that "*With a 10% adoption rate of EVs, time-of-day charging increased local renewable energy usage by 20% and enables marginal wind energy converters to upgrade*" [54]. Colmenar et al. proposed a novel grid technique in order to optimize the operation of RE and EVs to increase penetration of RE [54]. In our current research, RE, EDR, ER and EC are considered simultaneously. It is essential to highlight the importance of EC block which aim is to determine the energy structure generation as well as the RE contribution. Consequently, the EV owners know in advance, when it is the best moment to charge their vehicle based on an environmental point of view. This block can provide accurate forecasts owing to the implementation of neural networks. It

Table 4.5 Percentage of occupants belonging to each social group in Alcalá de Henares

Social group	Sample size (number of households)	Probability of success assessed by using the first sample	Probability of failure obtained when using the first sample	Sample size (number of households)	Probability of success when using the second sample	H_0 meaning	H_0 value ($\alpha = 0.01$)
Freelancers	150	0.15	0.85	100	0.16	The percentage of freelancers is close to 15%	Accepted
Commuters who used their vehicles		0.72	0.28		0.74	The percentage of commuters and local workers are close to 72% and 13% respectively	Accepted
Local workers		0.13	0.87		0.14		Accepted

Table 4.6 Percentage of occupants belonging to each social group in Jaén

Social group	Sample size (number of households)	Probability of success assessed by using the first sample	Probability of failure obtained when using the first sample	Sample size (number of households)	Probability of success when using the second sample	H_0 meaning	H_0 value ($\alpha = 0.01$)
Freelancers	150	0.33	0.77	100	0.35	The percentage of freelancers is close to 33%	Accepted
Commuters who used their vehicles		0.45	0.55		0.49	The percentage of commuters and local workers are close to 45% and 22% respectively	Accepted
Local workers		0.22	0.78		0.23		Accepted

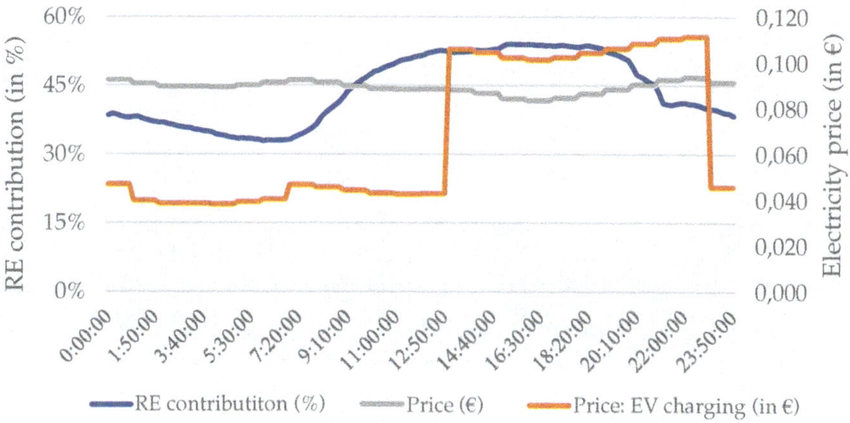

Fig. 4.10 RE contribution, electricity price and EV charging price. June 2020

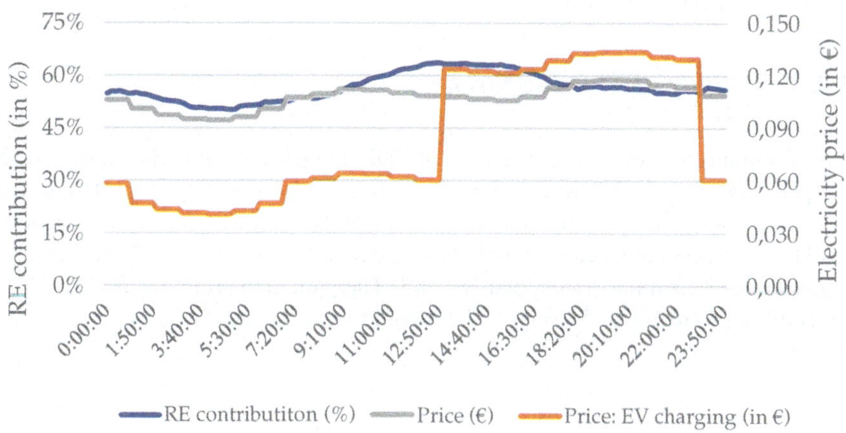

Fig. 4.11 RE contribution, electricity price and EV charging price. December 2020

must be stated that the network performance was assessed with an average value of 2.25E-7. Finally, the algorithm presented in this research uses the Open Charge Map® API which aims to inform the driver where the closest charging point is. Therefore, by using the information provided by this API, the energy consumption estimate assessed by the Here® API and the consumption models tuned properly, the driver can decide if it is possible to postpone the EV charging to a better slot in which the RE contribution is higher. Consequently, the Open Charge Map® API is also useful to make the charging process greener.

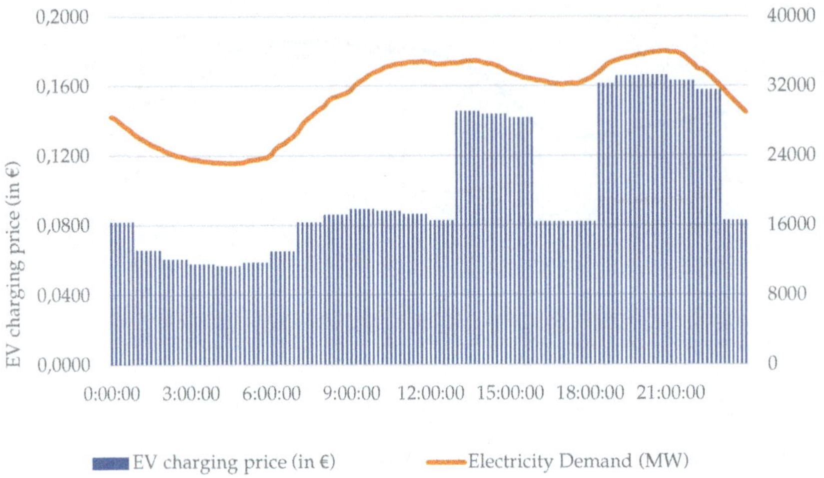

Fig. 4.12 EV charging price proposal

4.4.5 Validity of This Research

It is important to assess the contribution of EDR and ER to V2B taking into account the participation of EV owners. Several factors such as battery degradation and policies influence this participation. In this sensitivity analysis, several participations have been considered when the number of participants was between 25 and 75%. As Fig. 4.13 shows similar results that Fig. 4.8. The gain goes from 4.225 kWh (25% for Building F) to 106.775 kWh (75% for Building E) on a daily basis.

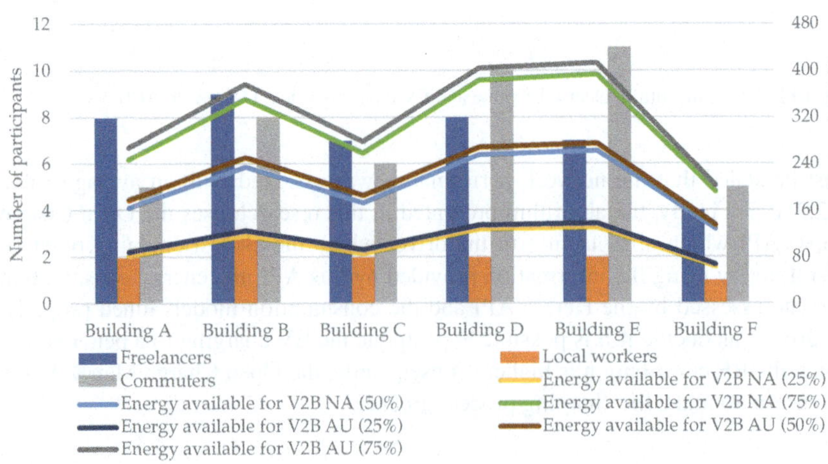

Fig. 4.13 V2B contribution

4.5 Conclusions

Emissions linked to the transport sector and building are of great concern nowadays. Consequently, improvement in both fields must be performed. This research proposed an algorithm based on the Here® application interface (one of the most important digital maps suppliers), neural networks, EVs, ER, EDR and EC concepts. By using this algorithm, the increase in energy available to be used in V2B technology was assessed. However, there is one important topic to analyse the energy available such as the stochastic usage of EVs. To be more specific, it is essential to classify the working population into social groups such as freelancers, local workers and commuters. Due to this, many data were collected in real-driving conditions from drivers who belonged to different social groups as their way of driving is different. Finally, all these data acquisitions were conducted in two cities (Alcalá de Henares—Madrid-Spain and Jaén-Spain) which are located in different climatic zones. The main conclusions that can be drawn are:

a. Energy savings

   This algorithm introduces reduction in energy consumption when driving EVs. As it could be expected, energy consumption is different depending on the social group. Consequently, the contribution to V2B technology differs. Regarding Alcala de Henares, energy efficiency reaches 2.2 kWh for freelancers per day. When it comes to commuters, this gain reaches 1.5 kWh a day and, finally, 0.6 kWh and for local workers on a daily basis. Regarding Jaén, the savings are similar. The energy efficiency reaches 1.9 kWh for freelancers per day. When it comes to commuters, this gain is 1 kWh on a daily basis and, finally, 0.6 kWh for local workers a day.

b. Contribution to vehicle-to-building

   V2B is based on the principle that the EV owner will participate and inject energy stored in the EV battery into the building. However, it is essential to determine energy available and, again, the fact of taking into account social groups influences energy available to be used for V2G technique. Regarding Alcalá de Henares, energy available ranges between 112 and 144 kWh a day depending on the social group mix existing in the building. In regard to Jaén, energy available ranges between 76 and 152 kWh a day depending on the social group mix existing in the building. Finally, it must also be taken into account that energy consumption pattern may change depending on the social groups that occupants belong to.

c. Charging policies

   In order to make the charging process greener, it is necessary to charge EVs when renewable energy contribution is higher. Based on the analysis of building consumption done in this research, the energy consumption pattern can differ depending on the social group that the occupant belongs to. The algorithm provided in this research can determine when the contribution of renewable energy is higher. Due to this, when its contribution is higher, the charging price is more expensive. This study proposes possible changes to charging fees to make V2B and V2G compatible. In order to apply this fee, the increase in megawatt of renewable energy installed is as important as increase in the number of EVs.

References

1. Economics for energy. https://eforenergy.org/docpublicaciones/informes/informe_transporte.pdf (accessed on 7 March 2021)
2. Spanish Ministry of Economics of https://www.miteco.gob.es/es/calidad-y-evaluacion-ambiental/publicaciones/02_16_transporte_pae2017_tcm30-498096.pdf (accessed on 7 March 2021)
3. Pillay, N.S.; Brent, A.C.; Musango, J.K. Affordability of battery electric vehicles based on disposable income and the impact on provincial residential electricity requirements in South -Africa. *Energy* 2019, 171–1087
4. Bastida-Molina, P.; Hurtado-Pérez, E.; Peñalvo-López, E.; Moros-Gómez, M.C. Assessing transport emissions reduction while increasing electric vehicles and renewable generation levels. *Transportation Research Part D: Transport and Environment* 2020, 88: 102560
5. Zheng, J.; Sun, X.; Jia, L.; Zhou, Y. Electric passenger vehicles sales and carbon dioxide emission reduction potential in China's leading markets. *Journal of Cleaner Production* 2020, 243: 118607
6. Helmers, E.; Dietz, J.; Weiss, M. Sensitivity analysis in the life-cycle assessment of electric vs. combustion engine cars under approximate real-world conditions. *Sustainability* 2020, 12: 1241
7. Messagie, M.; Boureima, F.S.; Coosemans, T.; Macharis, C.; Van Mierlo, J. A Range-Based Vehicle Life Cycle Assessment Incorporating Variability in the Environmental Assessment of Different Vehicle Technologies and Fuels. *Energies* 2014, 7(3): 1467–1482
8. Iwafune, Y.; Ogimoto, K.Economic Impacts of the Demand Response of Electric Vehicles Considering Battery Degradation. *Energies* 2020, 13(21): 5771
9. Vopava, J.; Koczwara, C.; Traupmann, A.; Kienberger, T. Investigating the Impact of E-Mobility on the Electrical Power Grid Using a Simplified Grid Modelling Approach. *Energies* 2019, 13(1): 39
10. Thompson, A.W.; Perez, Y. Vehicle-to-Everything (V2X) energy services, value streams, and regulatory policy implications. *Energy Policy* 2020, 137: 111136
11. Pearre, N.S.; Ribbering, H. Review of research on V2X technologies, strategies, and operations. *Renewable and Sustainable Energy Reviews* 2019, 105: 61–70
12. Bibak, B.; Tekiner-Mogulkock, H. A comprehensive analysis of Vehicle to Grid (V2G) systems and scholarly literature on the application of such systems. *Renewable Energy Focus* 2021, 36: 1–20
13. Bibak, B.; Tekiner-Mogulkock, H. Influences of vehicle to grid (V2G) on power grid: An analysis by considering associated stochastic parameters explicitly. *Sustainable Energy, Grids and Networks* 2021, 26: 100429
14. Parsons, G.R.; Hidrue, M.K.; Kempton, W.; Gardner, M.P. Willingness to pay for vehicle-to-grid (V2G) electric vehicles and their contract terms. *Energy Economics* 2014, 42: 313–324
15. Maeng, K.; Ko, S.; Shin, J.; Cho, Y. How Much Electricity Sharing Will Electric Vehicle Owners Allow from Their Battery? Incorporating Vehicle-to-Grid Technology and Electricity Generation Mix. *Energies* 2020, 13: 4248
16. Noel, L.; Zarazua de Rubens, G.; Kester, J.; Sovacool, B.K. Leveraging user-based innovation in vehicle-to-X and vehicle-to-grid adoption: A Nordic case study. *The Journal of Cleaner Production* 2021, 287: 125591
17. Geske, J.; Schumann, D. Willing to participate in vehicle-to-grid (V2G)? Why not!. *Energy Policy* 2018, 120: 392–401

18. Kester, J.; Noel, L.; Zarazua de Rubens, G.; Sovacool, B.K. Promoting Vehicle to Grid (V2G) in the Nordic region: Expert advice on policy mechanisms for accelerated diffusion. *Energy Policy* 2018, 116: 422–432
19. Mwasilu, F.; Justo, J.J.; Eun-Kyung, K.; Duo Do, T.; Jung, J.W. Electric vehicles and smart grid interaction: A review on vehicle to grid and renewable energy sources integration. *Renewable and Sustainable Energy Reviews* 2014, 34: 501–516
20. Rahbari, O.; Vafaeipour, M.; Omar, N.; Rosen, M.A.; Hegazy, O.; Timmermans, J.M.; Heibati, S.; Van DenBossche, P. An optimal versatile control approach for plug-in electric vehicles to integrate renewable energy sources and smart grids. *Energy* 2017, 134: 1053–1067
21. Mozafar, M.R.; Moradi, M.H.; Amini, M.H. A simultaneous approach for optimal allocation of renewable energy sources and electric vehicle charging stations in smart grids based on improved GA-PSO algorithm. *Sustainable Cities and Society* 2017, 32: 627–637
22. Gao, Y.; Jiang, J.; Zhang, C.; Zhang, W.; Jiang, Y. Aging mechanisms under different state-of-charge ranges and the multi-indicators system of state-of-health for lithium-ion battery with $Li(NiMnCo)O_2$ cathode. *Journal of Power Sources* 2018, 400: 641–651
23. Buonomano, A.; Calise, F.; Cappiello, F.L.; Palombo, A.; Vicidomini, M. Dynamic analysis of the integration of electric vehicles in efficient buildings fed by renewables. *Applied Energy* 2019, 245: 31–50
24. Zhou, Y.; Cao, S.; Hensen, J.L.M.; Lund, P.D. Energy integration and interaction between buildings and vehicles: A state-of-the-art review. *Renewable and Sustainable Energy Reviews* 2019, 114: 109337
25. Zhang, X.; Zou, Y.; Fan, J.; Guo, H. Usage pattern analysis of Beijing private electric vehicles based on real-world data. *Energy* 2019, 167: 1074–1085
26. Eco-driving behaviors of electric vehicle users: A survey study. *Transportation Research Part D: Transport and Environment* 2020, 78: 102188
27. Nunzio, G.; Thibault, L.; Sciarretta, A. A model-based eco-routing strategy for electric vehicles in large urban networks. In Proceedings of the International Conference on Intelligent Transportation Systems, Rio de Janeiro, Brazil, 1–4 November 2016, https://doi.org/10.1109/itsc.2016.7795927
28. Sureth, A.; Moll, V.; Nachtwei, J.; Franke, T. The golden rules of ecodriving? The effect of providing hybrid electric vehicle (HEV) drivers with a newly developed set of ecodriving-tips. *Transportation Research Part F: Traffic Psychology and Behaviour* 2019, 64: 565–581
29. Chakraborty, N; Mondal, A.; Mondal, S. Intelligent Charge Scheduling and Eco-Routing Mechanism for Electric Vehicles: A Multi-Objective Heuristic Approach. Sustainable Cities and Society 2021. In press
30. Orfila, O.; Saint Pierre, G.; Messias, M. An android based ecodriving assistance system to improve safety and efficiency of internal combustion engine passenger cars. *Transportation Research Part C: Emerging Technologies* 2015, 58: 772–782
31. Franke, T.; Georg Arend, M.; McIlroy, R.C.; Stanton, A. Ecodriving in hybrid electric vehicles – Exploring challenges for user-energy interaction. *Applied Ergonomics* 2016, 55: 33–45
32. Thijssen, R.; Hofman, T.; Ham, J. Ecodriving acceptance: An experimental study on anticipation behavior of truck drivers. *Transportation Research Part F: Traffic Psychology and Behaviour* 2014, 22: 249–260
33. Here® API https://developer.here.com/ (accessed on 7 March 2021)
34. Huebner, G.M.; Hamilton, I.; Chalabi, Z.; Shipworth, D.; Oreszczyn, T. Explaining domestic energy consumption—The comparative contribution of building factors, socio-demographics, behaviours and attitudes. *Applied Energy* 2015, 159: 589–600
35. Huebner, G.; Shipworth, D.; Hamilton, I.; Chalabi, Z.; Oreszczyn, T. Understanding electricity consumption: A comparative contribution of building factors, socio-demographics, appliances, behaviours and attitudes. *Applied Energy* 2016, 177: 692–702
36. Harputlugil, T.; de Wilde, P. The interaction between humans and buildings for energy efficiency: A critical review. *Energy Research & Social Science* 2021, 71: 101828
37. Pan, S.; Xiong, Y.; Han, Y.; Zhang, X.; Xia, L.; Wei, S.; Wu, J.; Han, M. A study on influential factors of occupant window-opening behavior in an office building in China. *Building and Environment* 2018, 133: 41–50

38. Yousefi, F.; Gholipour, Y.; Yan, W. A study of the impact of occupant behaviors on energy performance of building envelopes using occupants' data. *Energy and Buildings* 2017, 148: 182–198
39. European Commission https://setis.ec.europa.eu/sites/default/files/reports/Driving_and_parking_patterns_of_European_car_drivers-a_mobility_survey.pdf (accessed on 7 March 2021)
40. R Official documentation https://cran.r-project.org/web/packages/PASWR/index.html (accessed on 28 February 2021)
41. García-Pérez A. Estadística básica con R. 1st ed. Vizcaya: UNED; 2011
42. Elsevier. https://www.sciencedirect.com/topics/earth-and-planetary-sciences/kurtosis https://www.sciencedirect.com/topics/earth-and-planetary-sciences/kurtosis
43. Zhang, X.; Zou, Y.; Fan, J.; Guo; H. Usage pattern analysis of Beijing private electric vehicles based on real-world data. *Energy* 2019, 167: 1074–1085
44. Schey, S.; Scoffield, D.; Smart, J. A First Look at the Impact of Electric Vehicle Charging on the Electric Grid in The EV Project. *World Electric Vehicle Journal* 2012, 5(3): 667–678
45. Verichev, K.; Zamorano, M.; Carpio, M. Effects of climate change on variations in climatic zones and heating energy consumption of residential buildings in the southern Chile. *Energy & Buildings* 2020, 215: 109874
46. Li, M.; Cao, J.; Xionga, M.; Li, J.; Feng, X.; Meng, F. Different responses of cooling energy consumption in office buildings to climatic change in major climate zones of China. *Energy & Buildings* 2018, 173: 38–44
47. Li, J.; Yang, L.; Long, H. Climatic impacts on energy consumption: Intensive and extensive margins. Energy Economics 2018, 71: 332–343
48. Sheng, W.; Wen, B.; Zhang, L. Envelope performance of residential building in cool, warm and hot climatic zones: Results from self-designed in-situ monitoring campaigns. *Energy & Buildings* 2021, 232: 110655
49. Csoknyai, T.; Legardeur, J.; Abi Aklec, A.; Horvátha, M. Analysis of energy consumption profiles in residential buildings and impact assessment of a serious game on occupants' behavior. *Energy & Buildings* 2019, 196: 1–20
50. Kavousian, A.; Rajagopal, R.; Fischer, M. Determinants of residential electricity consumption: Using smart meter data to examine the effect of climate, building characteristics, appliance stock, and occupants' behavior. *Energy* 2013, 55: 184–194
51. Sagaria, S.; Neto, R.C.; Baptista, P. Modelling approach for assessing influential factors for EV energy performance. *Sustainable Energy Technologies and Assessments* 2021, 44: 100984.
52. Yan, J.; Zhang, J.; Liu, Y.; Lv, G.; Han, S.; Gonzalez-Alfonzo, I.E. EV charging load simulation and forecasting considering traffic jam and weather to support the integration of renewables and EVs. *Renewable Energy* 2020, 159: 623–641
53. Atia, R.; Yamada, N. More accurate sizing of renewable energy sources under high levels of electric vehicle integration. *Renewable Energy* 2015, 81: 918–925
54. Pearre, N.S.; Swan, L.G. Electric vehicle charging to support renewable energy integration in a capacity constrained electricity grid. *Energy Conversion and Management* 2016: 130–139

Chapter 5
Contribution of Driver Efficiency to the European Green Deal

Pedro-Miguel Ortega-Cabezas, Antonio Colmenar-Santos, David Borge-Diez, and Jorge-Juan Blanes-Peiró

Abbreviations

API	Application Programming Interface
EC	Eco-charging
EDR	Eco-driving
EGD	European Green Deal
ER	Eco-routing
EV	Electric Vehicle
GRU	Gated Recurrent Units
NAR	Nonlinear Autoregressive Neural Networks
RE	Renewable Energy
V2G	Vehicle-to-Grid
V2H	Vehicle-to-Home
VCU	Vehicle Control Unit

Extract from Energy, Can eco-routing, eco-driving and eco-charging contribute to the European Green Deal? Case Study: The City of Alcalá de Henares (Madrid, Spain), Volume 228, 1 August 2021, 120532. https://www.sciencedirect.com/journal/energy.

P.-M. Ortega-Cabezas · A. Colmenar-Santos (✉)
Departamento de Ingeniería Eléctrica, Electrónica y de Control, UNED, Juan del Rosal, 12, Ciudad Universitaria, 28040 Madrid, Spain
e-mail: acolmenar@ieec.uned.es

P.-M. Ortega-Cabezas
e-mail: pedro-miguel.ortega-cabezas@valeo.com

D. Borge-Diez · J.-J. Blanes-Peiró
Department of Electrical and Control Engineering, Universidad de León, Campus de Vegazana, s/n, 24071 León, Spain
e-mail: david.borge@unileon.es; dbord@unileon.es

J.-J. Blanes-Peiró
e-mail: jorge.blanes@unileon.es

© The Author(s), under exclusive license to Springer Nature Switzerland AG 2024

A. Colmenar-Santos et al. (eds.), *Development and Testing of Vehicle Software and its Influence on Sustainable Transport*, https://doi.org/10.1007/978-3-031-47630-3_5

5.1 Introduction

The European Green Deal (EGD) aims to make Europe the first climate-neutral continent as well as to reduce emissions of greenhouse gases by 2050. This research offers proposals for this deal based on sustainable transport, clean energy and reduction in energy consumption of the buildings. An algorithm based on Here® application programming interface, neural networks, data from the Spanish transmission system operator, eco-routing (ER), eco-driving (EDR) and eco-charging (EC) is proposed. Its contribution to vehicle-to-home (V2H), renewable energy (RE) integration, V2H and vehicle-to-grid (V2G) compatibility is analyzed by using data acquisitions of trips made by drivers belonging to different social groups.

The algorithm allows saving per day up to 2.2 kWh for freelancers, 1.5 kWh for commuters and 0.6 kWh for local workers. The V2H contribution is increased from 18 to 553 MWh per year. Finally, neural networks allow a better integration of RE.

The contributions to the EGD are the following. Energy efficiency is improved when policies are addressed to the adequate social sector combined with EDR and ER. Neural networks allow achieving a better integration of renewable energies as they predict when its contribution is higher. Policies to make V2G and V2H compatible to reduce emissions must be developed.

The EGD aims to make Europe the first climate-neutral continent. EGD is focused on three pillars: reduction in greenhouse emissions by 2050, stimulation of economic growth without linking it to resource use and involvement of all members of society in the implementation of this new strategy [1]. To achieve the EGD goals many actions must be undertaken by all sectors of the economy, including: "*investing in environmentally-friendly technologies, supporting industry to innovate, rolling out cleaner, cheaper and healthier forms of private and public transport, decarbonizing the energy sector, ensuring buildings are more energy efficient and working with international partners to improve global environmental standards*" [2, 3]. This research is focused on these concepts: sustainable mobility, reduction in energy consumption of buildings and clean energy. Regarding sustainable mobility, the EGD aims to reduce 90% of greenhouse gas emissions, smarter traffic management, and the promotion of other modes of transport. Considering these initiatives, it is vital to promote electric vehicles (EVs). Regarding the clean energy concept, the EGD aims to increase the presence of RE in the electricity sector. As stated by the European Union, "*prioritize energy efficiency and develop a power sector based largely on renewable energy sources*" [4]. Finally, the goal of the EGD about improving the energy efficiency of buildings is to prioritize the "*design and consumption of new retrofitting of existing buildings as zero-emission/zero pollution, positive energy powerhouses with sustainable green neighborhoods*" [1]. One of the major components of this transition is energy positive buildings with sustainable and RE technologies.

Road transportation accounts for 73% of the total demand in the transport sector in 2017, which means 34% higher than in 1990 [5]. EVs play a key role in decarbonizing the transport sector and in achieving the goal of reducing greenhouse gas emissions by 80% by 2050 [6]. Even if the whole life cycle is considered, the sum of emissions from all associated EV activities (production, maintenance and end-of-life stage) is lower than the ones of traditional vehicles despite the battery production process [7–9]. Considering EV prices, new research shows that EVs are already cheaper than diesel or petrol ones as shown by Carrington [10] who has analyzed several factors such as costs over four years, purchase price, fuel, insurance, taxation and maintenance. Colmenar et al. [11] prove how the right policies are essential for widespread use of EVs by analysing the profitability of an electric transportation model.

V2G technology allows a better integration of RE and energy peak reduction [12, 13]. Hofmann et al. [14] show that EVs cause zero pollution as the electricity sector is decarbonized in China. This conclusion is supported by Thiel et al. [15]. V2H allows reducing emissions linked to buildings as EVs' owners can supply their homes with the energy stored in EV batteries. Colmenar et al. [16] show the benefits of V2H as it increases the penetration of REs and the reduction of costs. Noori et al. [17] focused on how V2H contributes to achieving the energy requirements for net zero energy buildings. By combining the best design alternatives which make buildings energy efficient, solar photovoltaic sources that meet the remaining energy demand and V2H, grid electricity consumption can be reduced by 68% compared to that of a conventional building design. V2H and V2G have also some drawbacks, as the supplementary use of the battery implies that the battery service life is reduced. Darcovich et al. [18] proved that V2H technique provides useful services with acceptable battery degradation in certain conditions. Lazzeroni et al. developed an approach to investigate EV battery management aiming to minimize electricity costs based on the driver's behaviour and battery constraints [19].

EDR minimizes fuel consumption and emissions as drivers are informed about how efficient their method of driving is [20]. Qi et al. [21] did research into EDR by quantifying the energy saved when applying to EVs. Sabrina et al. [22] published a study in which continuous and on-demand feedback on driving behaviour and safety system is performed. Zhan et al. [23] show how energy efficiency is enhanced thanks to the systems in charge of monitoring the EV battery. ER intends to plan the most efficient route to go from point *A* to *B* considering several parameters such as the real-time traffic condition, road types, the number of passengers and cargo weight [24, 25]. Nunzio et al. [26] propose a model which considers speed fluctuations and road network infrastructure to set the route. Some systems based on collecting energy consumption data in-real word driving conditions to set the optimal route have been offered [27]. ER, EDR and EC reduce emissions and improve energy efficiency, as they reduce energy consumption and contribute to better integration of RE in the EV lifecycle [28, 29].

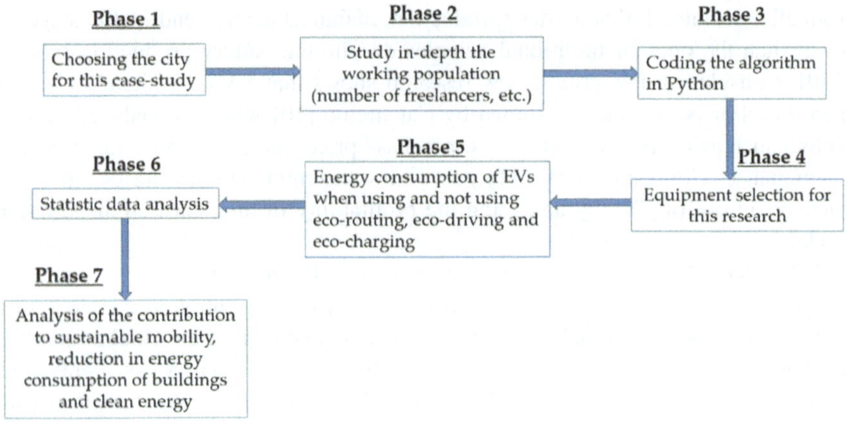

Fig. 5.1 Method description

5.2 Methods

5.2.1 Description

Two key points must be considered in the method used in this research: the analysis of the working population of the city subjected to this case-study and the algorithm proposed. The former consists of classifying its population into different social groups such as freelancers, commuters and local workers as the way of using EVs is different. Then, data acquisitions by using and not using the algorithm were done considering the usage of EVs by the drivers participating in this research. All trips taken in this study were made according to the users' needs (aleatory trips/not previously planned). The latter was coded in Python by using the Here® API [30], being able to determine the best route considering EDR and ER models built by analysing data provided by the vehicle control unit (VCU), traffic conditions, drivers' habits, potential recharge needs among others [30, 31]. The EC concept shows drivers when a recharge process should be performed considering when RE contribution is higher (Fig. 5.1).

5.2.2 Case Study Selection Criteria

Alcalá de Henares is a city located in the Community of Madrid at a distance of 32 km from the capital of Spain. According to the data published by the Spanish National Institute of Statistics, in 2018, it had a total of 198,750 inhabitants. The traffic is monitored by using 17 cameras distributed across the whole city (Fig. 5.2) [32, 33]. For the extension of the city (marked in black), the cameras are mainly concentrated in the South (Fig. 5.2). In Fig. 5.3 the main road access directions of the capital of Spain are marked. They are highly conflicting points in matters of traffic density [32, 33].

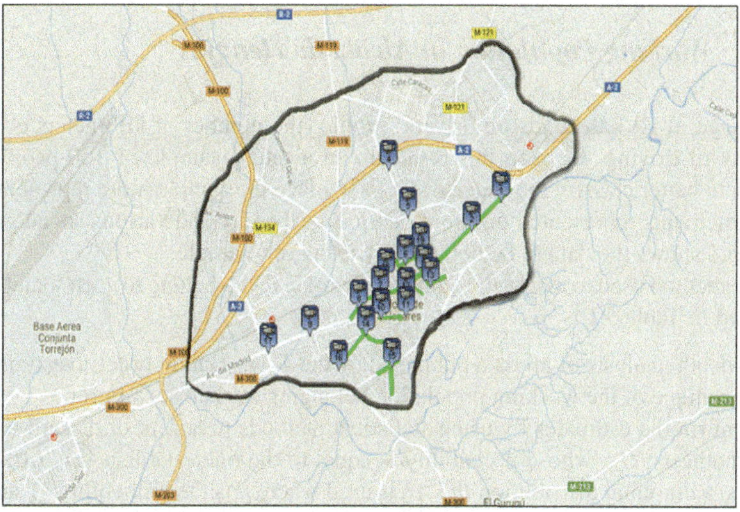

Fig. 5.2 Cameras distribution

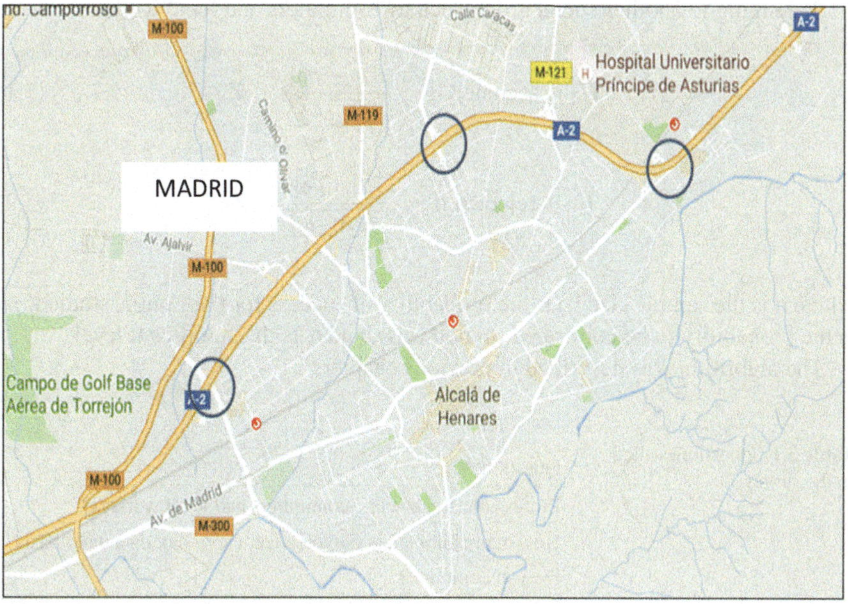

Fig. 5.3 Road access direction Madrid

Considering the population, the size of the city (the second biggest city in the Community of Madrid), traffic conditions and location, the authors considered this city as a good candidate to perform this research.

5.2.3 Working Population in Alcala de Henares

The usage of EVs depends on drivers' profile (the number of kilometers covered, the way of driving, etc.). In this research, an aleatory sample of 100 people was chosen to better classify the usage of EVs. To make the sample more representative, the participants represented different districts of the city and various social groups. Table 5.1 shows the data to be obtained by using this sample.

The method used to assess the number of workers who belong to each social group depicted in Table 5.1 is:

a. In-depth analysis of all data published by the Town Hall in order to estimate the percentage of the working population belonging to each social group.
b. Confirm the estimates by using statistical methods to accept or discard the null hypothesis, H_0, whose probability is equal to the one established in (a). To do this, a binomial distribution $B(1, p)$ is used where p is the probability of success. The H_0 values are analyzed in the result section by using Eqs. (5.1) and (5.2) [34]. For example, regarding the freelancer group, p is the probability of choosing a freelancer and $1 - p$ is the probability of choosing a member of the sample belonging to another social group such as commuters and local workers.

$$H_0 \text{ is accepted if } \quad \frac{|\hat{p} - p_0|}{\sqrt{\frac{p_0(1-p_0)}{n}}} \leq Z_{\frac{\alpha}{2}} \tag{5.1}$$

$$H_0 \text{ is rejected if } \quad \frac{|\hat{p} - p_0|}{\sqrt{\frac{p_0(1-p_0)}{n}}} > Z_{\frac{\alpha}{2}} \tag{5.2}$$

where n is the sample size, $\hat{p}$ is the probability of success for the sample studied, p_0 is the probability to be confirmed (hypothesis), and α is the significant level.

The algorithm was described in previous chapters.

Table 5.1 Questions asked in the survey

Questions
Profession (freelancers, commuters and local workers)
How many kilometers do you cover every day on a daily basis?
Do you commute?
Do you use public transport?
Are you willing to buy an EV?

5.3 Results

5.3.1 *Number of People of Each Social Group*

A statistical method described later was followed to determine the number of people of each social group.

a. Freelancers

Freelancers usually cover a great deal of kilometers on a daily basis. Considering the data provided by the Town Hall, the number of freelancers' trend in Alcalá is shown in Fig. 5.4 and Table 5.2 [35, 36]. Table 5.2 depicts the number of freelancers classified by age range. Most of them are included in the range between 35 and 54 years old. In the last two years, the number of freelancers is very close to 14.5% following a very stable trend (Fig. 5.4). The maximum percentage is slightly higher than 16% in 2015 and the minimum value is 14.2% in 2016. The last known value was 14.6%. In this research, the percentage of freelancers used is 14.6%. This value is confirmed statically. A 100 sample was done with the aim of assessing the number of people belonging to each social group considering the working population (freelancers, commuters and local workers). The number of freelances found in the sample was 15 people. All of them use a vehicle to get to work. Therefore, the hypothesis can be confirmed, and the percentage of freelancers considered in this research is 14.6% by using Eqs. (5.1) and (5.2) described earlier [34].

b. Workers who commute and do not commute

Table 5.3 depicts the number of people belonging to different social groups considered as the initial hypothesis. These values were established as baseline for the sensitivity analysis. Regarding freelancers, the initial hypothesis was established by using the statistical data published by the Town Hall (Table 5.2 and Fig. 5.4). Regarding

Fig. 5.4 Freelancer trend in Alcalá

Table 5.2 Number of freelancers

Age	Total workers	Total Freelancers
20–24	9,927	1,449
25–29	10,900	1,591
30–34	12,514	1,827
35–39	15,016	2,192
40–44	17,813	2,601
45–49	16,774	2,449
50–54	14,550	2,124
55–59	12,590	1,838
60–64	11,709	1,710
	121,793	17,181

commuters who use their own vehicles and local workers, there is no statistical data published by the Town Hall. Consequently, a first sample consisting of one hundred participants provides information about the number of people commuting (47%) and local workers (11.7%). Regarding the number of commuters who use the public transport, estimates show that 7,147,000 people (32,486 people on a daily basis) use trains to commute to their jobs [35, 36]. These workers are supposed not to have EVs as they are only used to reach the proximity of the station. Consequently, this research does not consider potential EVs belonging to this group of workers. Considering the questions shown in Table 5.1, the results show that only 2% of the sample of people who use public transport would be willing to buy EVs.

The sample consisting of one hundred people was also used to confirm the number of freelancers by using the H_0 value obtained from Eqs. (5.1) and (5.2). An additional sample of 50 individuals was used to confirm the number of local workers and commuters by employing again Eqs. (5.1) and (5.2). Table 5.4 shows how the acceptance of the percentage of each social group was conducted.

Table 5.3 Number of workers

	Number of workers	Number of workers (in %)
Freelancers	17,782	14.6
Commuters who use public transport	32,486	26.7
Commuters who use their vehicles	57,220	47
Local workers	14,305	11.7
Total number of workers	121,793	

Table 5.4 Percentage of each social group

Social group	How initial hypothesis is obtained	Probability of success obtained considering the 100-worker sample	Probability of failure obtained considering the 100-worker sample	Initial hypothesis in %	Sample used to confirm the hypothesis	Probability of success when using the second sample (50 participants)	H_0 meaning	H_0 value ($\alpha = 0.01$)
Freelancers	By using the statistical data published by the Town Hall	0.15	0.85	14.6	100	Not necessary	The percentage of freelancers is close to 14.6%	Accepted
Commuters who used their vehicles	By using a sample of 100 individuals as no statistical data are available	0.45	0.55	47	50	0.5	The percentage of commuters and local workers are close to 47% and 12% respectively	Accepted
Local workers		0.13	0.87	11.7		0.11		Accepted
Commuters who used public transport		0.27	0.73	26.7		0.26		Accepted

Table 5.5 EV consumption in kWh

Factor	Freelancers		Commuters		Local workers	
	A.U.[a]	N.A.[b]	A.U.[a]	N.A.[b]	A.U.[c]	N.A.[d]
Mean	24	26.2	9	10.5	3.3	3.9
Std deviation	0.6	0.4	0.3	0.32	0.32	0.29
Kurtosis	3.7	4.0	3.7	4.5	4.1	4.2
Skewness	−0.135	−0.121	−0.041	−0.032	−0.025	−0.015
p-value	0.395	0.401	0.401	0.415	0.396	0.399

[a]N.A. means no algorithm is used
[b]A.U. means the algorithm is used

5.3.2 Consumption for Each Social Group

The data collected from freelancers, commuters and local workers help to estimate statistically the EV consumption, and the improvement in energy efficiency obtained thanks to the algorithm. A lot of measurements were performed considering different drivers' profiles such as freelancers and local workers and commuters. All these measurements must be analyzed statistically to assess the average EV consumption for each social group. Table 5.5 depicts the statistical results for each social group obtained when analyzing the consumption data during the trips performed. The most important parameters to be analyzed is skewness, kurtosis and *p*-value. Regarding freelancers, skewness is almost zero. Therefore, the distribution is symmetric. Kurtosis values prove that the data distribution tails do not differ from normal distribution ones. The *p*-value represents the null hypothesis: the data follow a normal distribution. The null hypothesis can be considered as True if $p\text{-value} > 0.05$. Considering that the Skewness and the Kurtosis are sensitive to the sample size, the normality test was also confirmed by employing Q-Q plots and histogram. According to the results obtained, the hypothesis H_0 is confirmed. The number of kilometers ranges from 95 to 110 km per day. Regarding local workers and commuters, the analysis is similar. The *p*-value confirms the hypothesis H_0. This conclusion has also been drawn by using the graphical method such as Q-Q plots, box plots and histograms. The number of kilometers ranges from 60 to 75 km per day, and 3 km to 5 km per day for commuters and local workers respectively.

5.3.3 Sensitivity Analysis

EGD aims to improve energy efficiency of buildings. The impact of ER, EDR and EC in V2H to enhance energy consumption of buildings is analyzed. The trend of EV presence in Alcala in the years to come makes it necessary to estimate available energy for V2H. Table 5.6 and Fig. 5.11 show some potential scenarios. In Fig. 5.5, series entitled "*sales scenario*" represents the increasing number of EVs present in Alcala according to Table 5.6. Series "*EV presence*" represents the annual ratio (in percentage) of the number of EVs to the cark park trend in Alcala [37]. Regarding the sale hypothesis, this assumption is justified by several international estimations. EV sales have been increasing for the last years and different institutions forecast that this trend will continue in the future [38]. Other institutions confirm this trend and estimate that more than 18 million vehicles will be on the road in 2030 [39]. These statements are also based on the fact that the battery cost will be reduced by 2030 as discussed by the International Council on Clean Transportation [40]. After having analyzed the historical data of Alcala's cark park, one can conclude that:

a. Except for 2008, the car park is always below 100,000 vehicles.
b. Considering the last decade, the average car park was 95,298.

Consequently, it has been considered a 0.2% increment every year. In case of an error in this prediction, several scenarios have been considered.

The energy saving obtained earlier and Fig. 5.5 will be used to assess the energy available for V2H.

Table 5.6 Hypothesis used in this research

Scenario	Sale hypothesis	Trend of Alcalá de Henares car park
Hypothesis 1	An estimate of 35% increase in EV sales	1.002% increment
Hypothesis 2	An estimate of 10% increase in EV sales	1.002% increment
Hypothesis 3	No increase in EV sales occurs	1.002% increment

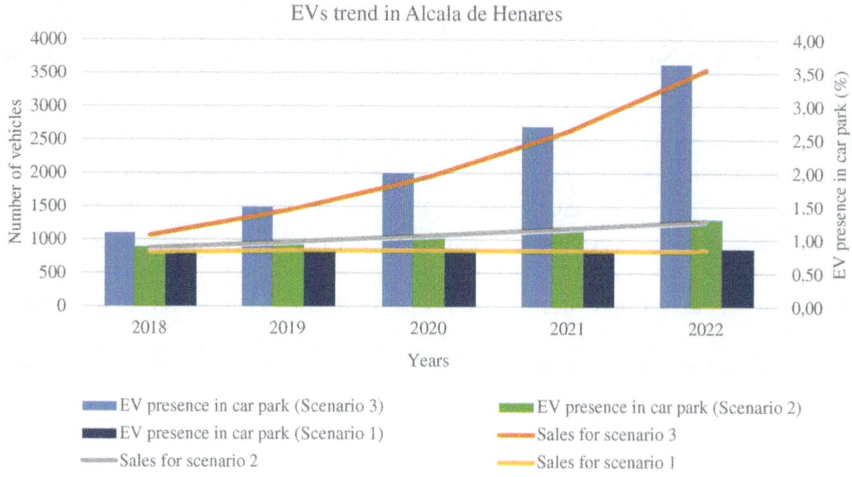

Fig. 5.5 EVs trend

The trend of the number of freelancers, commuters and local workers is linked to the economic climate in the year to come. Potential scenarios are completely influenced by the current pandemic. Such measures as employment regulation plans and different economic aids have been introduced by Governments to reduce the Covid-19 impact to keep companies from cutting jobs. Unfortunately, some companies unable to overcome this health crisis in the short-term will disappear. In the medium and long-term, the economic recovery is expected even if some disagreements are expressed regarding the recovery speed [40]. Three scenarios (Table 5.7) have been enacted considering the predictions made by the Spanish National Bank [40].

- *Scenario 1*

The number of EVs is represented by the series entitled "*Sales for scenario 1*" in Fig. 5.5. The number of working people in the city is reduced by five percent for all social groups in 2020. This percentage is based on different estimate in which the unemployment rate will be increased by 21.7% in 2021 according to the Spanish National Bank [41]. In 2021 and 2022, the unemployment rate will be reduced by 2% according to the estimates made by the same institution in such a way that in 2022 the figures of freelancers, commuter and local workers will be similar to those in 2020. The percentage of workers who belong to each social group and the number of EVs are indicated in the baseline column (Table 5.7). This scenario is linked to an increase in unemployment rate and EV sales stagnation.

- *Scenario 2*

The second scenario tries to describe the impact of Covid-19 and a more optimistic increase in working population. The Spanish National Bank estimates that even if the working population is decreased in 2020, the unemployment rate will decrease during 2021 and 2022. The authors have considered a more optimistic scenario than the Spanish National Bank [41]. Consequently, the unemployment rate will be

Table 5.7 Scenarios for assessing energy available for V2H

Factors	Baseline	Scenario 1			Scenario 2			Scenario 3		
		2020	2021	2022	2020	2021	2022	2020	2021	2022
Number of EVs	850	850	850	850	1,058	1,164	1,280	1,956	2,641	3,565
Number of freelancers	17,782	16,893	17,230	17,575	16,893	17,568	18,270	16,893	17,906	18,980
Number of commuters who use their vehicles	57,220	54,359	55,446	56,555	54,359	56,533	58,794	54,359	57,620	61,077
Number of local workers	14,305	13,590	13,862	14,139	13,590	14,133	14,698	13,590	14,405	15,269

reduced by 4% percent for all social groups in 2021 and 2022. The number of EVs is represented by the series entitled "*Sales for scenario 2*" in Fig. 5.5.

- *Scenario 3*

The third one assesses the number of workers belonging to each social group when a significant enhancement of economic situation takes place (an increase of six percent in working population). The number of EVs sold is represented by the series entitled "*Sales for scenario 3*" in Fig. 5.5.

Figure 5.6 shows the results from EDR, ED and EC to V2H obtained for freelancers considering each scenario. ER and EDR enhance energy efficiency which ranges between 82 MWh (scenario 1) and 343 MWh (scenario 3) per year. Regarding commuters (Fig. 5.7), energy efficiency ranges between 132 MWh (scenario 1) and 553 MWh (scenario 3) per year thanks to EDR and ER. Finally, Fig. 5.8 depicts the results when it comes to local workers. In this case, the energy efficiency enhancement ranges from 18 MWh (scenario 1) to 75 MWh (scenario 3) per year. These results show that not only the number of EVs is essential to increase energy available for V2H but the social group in which the presence of EVs is higher. Considering Figs. 5.6, 5.7 and 5.8, EDR and ER algorithm can contribute in a very important way to increment of the energy available for freelancers and commuters to V2H. When it comes to local workers, EDR and ER have not a significant impact as the average mileage and driving time are not as high as for freelancers. It must be stated that the number of EVs for each social group was assigned considering the values indicated in Table 5.3.

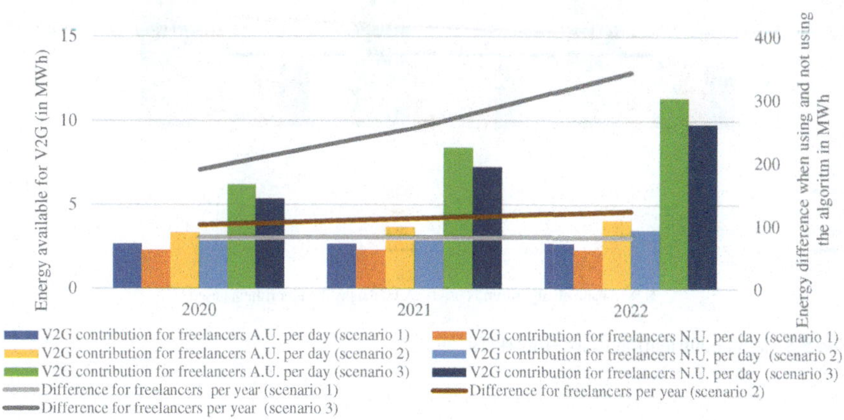

Fig. 5.6 V2G contribution to freelancers

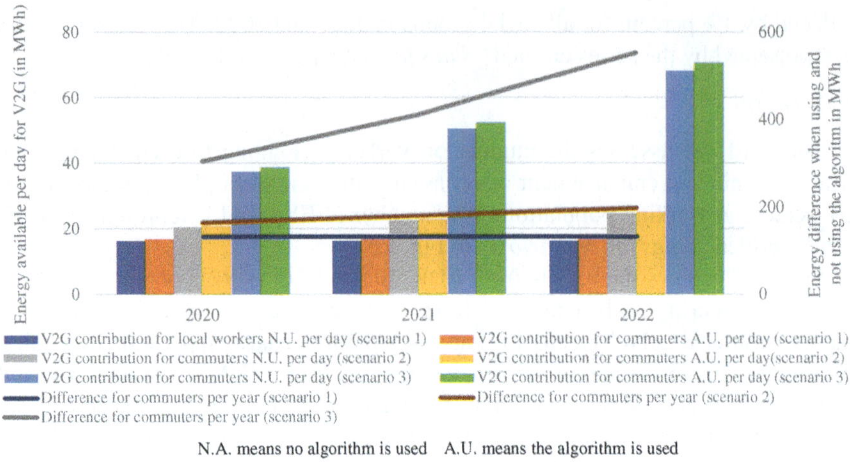

Fig. 5.7 V2G contribution to commuters

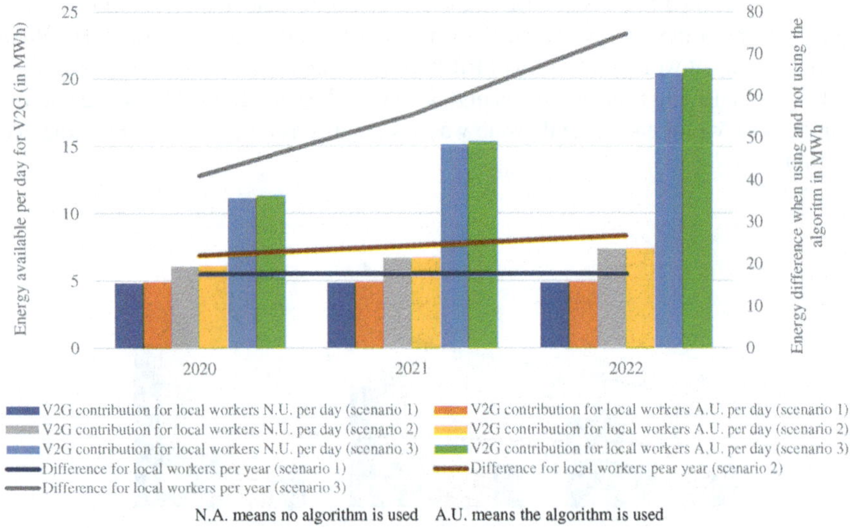

Fig. 5.8 V2G contribution to local workers

5.4 Discussion

5.4.1 *Sustainable Transport and Energy Consumption of Buildings*

During this research, EDR and ER have been used to prove that energy efficiency is enhanced considering social groups. The fact of improving energy efficiency implies that the energy consumption of buildings will be reduced thanks to V2H. The results show that:

a. Commuters

Commuters get frequently stuck in traffic jams as the traffic situation is more complicated when leaving big and medium size cities. The contribution to V2H when using ER and EDR is significant for this social group (Fig. 5.7). The energy saving is increased as the time spent by drivers in traffic jams is reduced. An increase ranging between 132 and 553 MWh per year occurs without enhancement in battery capacity. Thus, policies which encourage to buy EVs focusing on commuters can contribute to reduction of local emissions and increase energy efficiency in a more important way thanks to ER and EDR (Fig. 5.7). In Sect. 5.4.3, it is discussed whether commuters and freelancers are expected to participate in V2H and V2G at the same time.

b. Local workers

Local workers do not cover a lot of kilometers on a daily basis. The difference between using and not using EDR and ER does not have a significant impact on energy efficiency. Without considering ER and EDR, the energy available for V2H is not high (Fig. 5.8) as the number of local workers is low comparing to commuters and freelancers. Two conclusions can be drawn. Firstly, enhancements in energy available for V2H must be focused on increasing battery capacity as well as reducing harm due to charging and discharging processes as ER and EDR do not have an important contribution to energy efficiency. Secondly, drivers can participate in V2G and V2H as they do not cover many kilometers.

c. Freelancers

Freelancers represent the social group of people who drive more kilometers in mixed traffic conditions in a more significant way. Consequently, the results are between the ones obtained for commuters and local workers. ER and EDR allow increasing energy saving and the amount of energy available for V2H.

Energy efficiency is improved if policies to increment EV sales are addressed to the adequate social groups combined with ER, EDR and EC. If policies encourage commuters and freelancers to buy EVs, energy saving is increased when using ER and EDR. Therefore, ER and EDR combined with right policies contribute to the achievement of the goals indicated in the EGD.

5.4.2 Clean Energy

EGD aims to decarbonize the electricity sector by promoting RE. The number of RE MW must increase according to electricity consumption. Otherwise, investments in RE facilities will not be justified. Policies should encourage charging when RE reaches the highest contribution (from 12 a.m. to 6 p.m.) to reduce emissions at night (from 8 p.m.) when the second consumption peak takes place. However, this is not possible as the price for charging batteries is expensive when RE contribution is high (Figs. 5.9 and 5.10). The main conclusion is that promotion of RE and policy prices are as important as policies to promote EVs penetration into the market or charging points implementation. Regarding RE in Spain, the number of MW available is almost stable since 2012. Consequently, this topic is a key concern for the promotion of EVs in V2H and V2G in Spain. All aforementioned details are shown in Figs. 5.9 and 5.10 which represent EV charging price, the electricity price in the market and the RE contribution in %.

To accomplish better integration of RE, the EC block plays an essential role as it can inform the driver when it is the best moment to charge the EV battery considering the RE contribution. Most drivers decided to charge at night instead of the time when RE contribution was high despite the EC block indications (Fig. 5.11) as it is more. Thus, the prices set for charging EVs should be coherent with RE contribution. Likewise, RE installed could be increased as the consumption between 12 a.m. to 6 p.m. (when RE contribution is higher) is also increased.

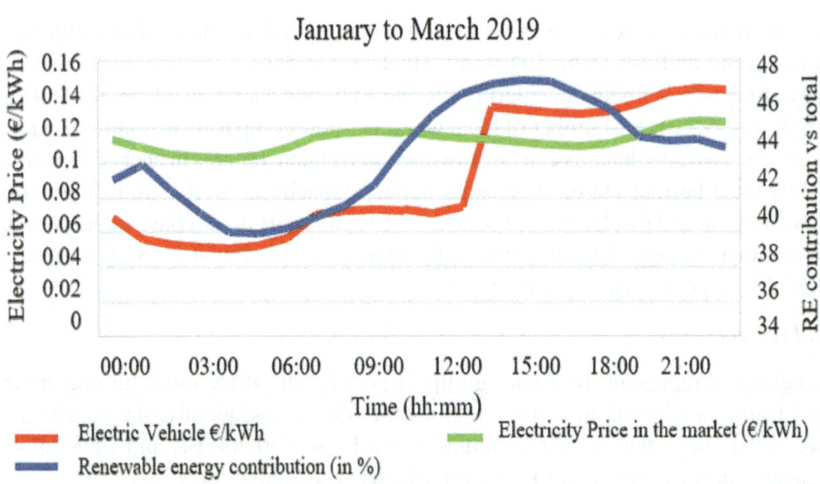

Fig. 5.9 Electricity prices trend

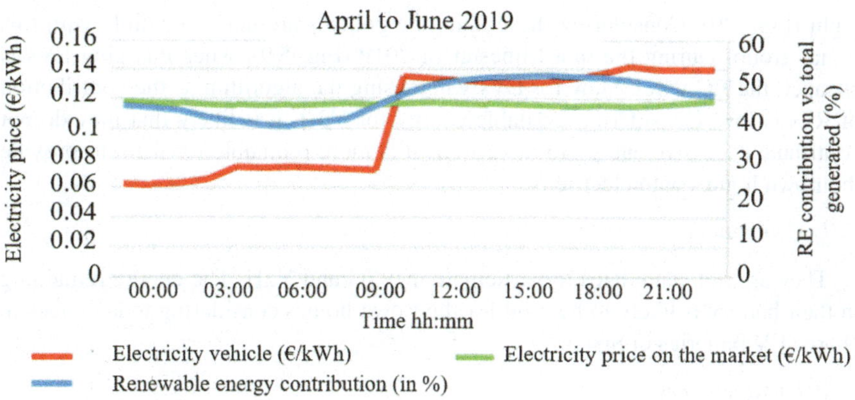

Fig. 5.10 Electricity prices trend

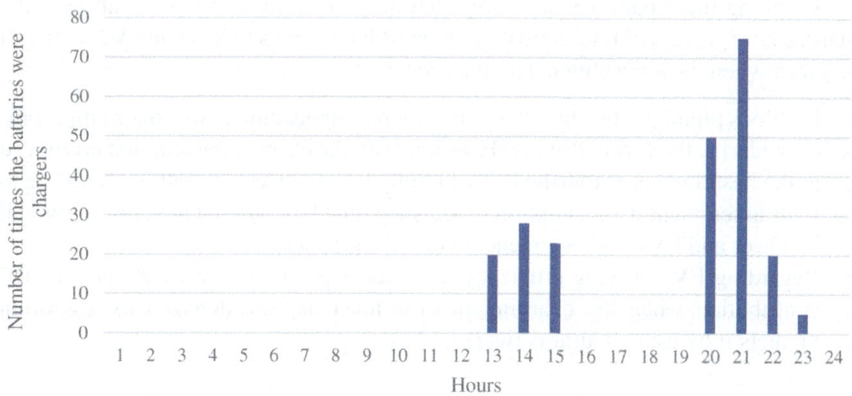

Fig. 5.11 Number of times the batteries were charged in this study

5.4.3 V2H and V2G Compatibility

Policies do not guarantee the compatibility of V2G and V2H considering the price to recharge EVs even if EVs were more promoted and the sales increased. Regarding each social group:

a. Freelancers

Its contribution to V2H and V2G will be low unless they charge batteries during the day before going back home. The relation between the electricity price on the market, special fees for charging EVs and RE contribution must be analyzed (Figs. 5.9 and 5.10). When the contribution of RE is higher, charging EVs is expensive. This happens from 12 a.m. as RE contribution is high but charging EV is expensive. The current fee established by the Spanish Government encourages recharging EVs at night (Fig. 5.9). Considering these data, the policies are not compatible with this social group. During the first trimester of 2019 (Fig. 5.9), when charging EVs is cheaper, the EC will be low for EVs when using the algorithm as the contribution of REs is low. Considering available energy for V2H, it is likely that they do not contribute excessive energy to the grid as it is more profitable for them to provide their own homes with electricity.

b. Commuters

They are not candidates to participate in V2G and V2H. The energy remaining in their battery is likely to be used for their own homes considering today's fees to charge EV batteries in Spain.

c. Local workers

The only candidates to participate in V2H and V2G are the ones who work in Alcalá. However, this social group is far from being the most representative one.

Assuming that Spain reaches a high RE mix and a high EVs penetration in the market, how green will EVs contribution be when it comes to V2H and V2G? It will be green when the algorithm enters the scene:

a. EC block predicts when the RE contribution is higher aiming to reduce emissions.
b. Considering the way of driving (EDR) and ER, the energy consumption is reduced as the best route is established considering the energy consumption according to EDR assessment (Consumption models are tuned). Consequently, pollution will be lower and EVs will be greener and eco-friendlier.
c. Regarding EVs, driving efficiency does not depend only on EDR and ER. RE contribution when the charging process must be considered. This algorithm predicts it by using neural networks.

5.4.4 Neural Network Validation

Before performing the trips, the following verifications were done on a daily basis. A comparison between the results published by the transmission system operator and the ones obtained when using the neural network was made. Figure 5.12 the performance of the GRU networks. All neural networks were implemented from the data of the transmission system operator the day before. This procedure assures that neural networks predict and work properly.

5.5 Conclusions

This research was conducted in the city of Alcalá de Henares (Madrid-Spain) aiming to establish the impact of EDR, ER and EC concepts on several topics linked to the EGD. To do this, an algorithm coded in Python is proposed based on the Here®

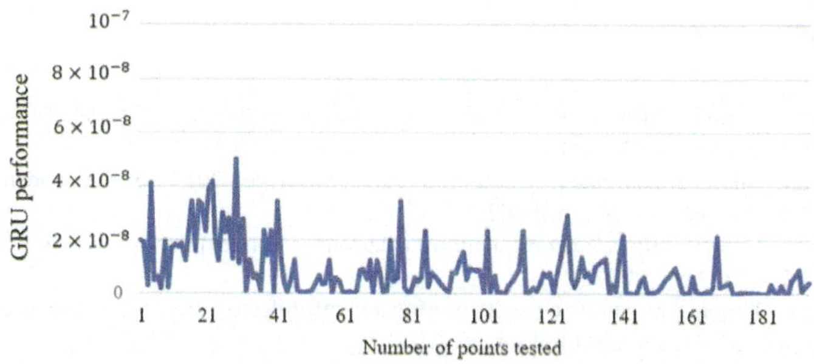

Fig. 5.12 Performance of a GRU network

programming application interface, neural networks and the data published by the Spanish system operator to reduce energy consumption and to choose the optimal moment to charge the battery. Data acquisitions were performed considering drivers' profile taking into account the most important social groups existing in the working population such as freelancers, local workers and commuters. The results obtained show that EDR, ER and EC can save a significant amount of energy in the future if policies change in Spain. To be more specific, the results show that energy efficiency per vehicle and per day could reach 1.5 kWh per day for commuters, 0.6 kWh per day for local workers and 2.2 kWh for freelancers when using the algorithm.

Several contributions to the EGD can be made:

a. It is vital to address policies to some social groups (freelancers, commuters, etc.) as not all of them contribute in the same way.
b. The energy efficiency obtained when using the algorithm increases the energy available for V2H. The amount of energy available for V2H is increased from 18 to 553 MWh depending on the social group.
c. The EC block implemented in this research allows drivers to charge EVs when renewable contribution is high. Consequently, renewable energies are better integrated to decarbonize the electricity sector.
d. Freelancers and commuters who cover a lot of kilometers on a daily basis do not usually participate in V2H or V2G unless policies establish lower prices when RE contribution is high. Nowadays in Spain, prices are higher when renewable contribution is high.
e. Policies cannot change if the power of RE installed in Spain is not increased. As detailed in (c), part of the demand could be moved when the RE contribution is high thanks to the EC block implemented in this algorithm.
f. The results show that current policies establishing the cost of EV charging in Spain are not compatible with high contribution to V2H and V2G for different groups of society.

References

1. European Union. https://eur-lex.europa.eu/legal-content/EN/TXT/?qid=1576150542719&uri=COM%3A2019%3A640%3AFIN; 2019 (accessed on 20 April 2022)
2. European Union. https://ec.europa.eu/info/strategy/priorities-2019-2024/european-green-deal_en; 2019 (accessed on 20 April 2022)
3. European Union. https://ec.europa.eu/info/files/annex-roadmap-and-key-actions_en; 2019 (accessed on 20 April 2022)
4. European Union. https://ec.europa.eu/info/sites/info/files/european-green-deal-communication_en.pdf; 2019 (accessed on 20 April 2022)
5. European Union. https://www.eea.europa.eu/data-and-maps/indicators/greenhouse-gas-emission-trends-6/assessment-3; 2019 (accessed on 15 June 2022)
6. European Environment Agency. https://www.eea.europa.eu/publications#c7=en&c11=5&c14=&c12=&b_start=0&c13=Electric%20vehicles%20and%20the%20energy%20sector%20-%20impacts%20on%20Europe's%20future%20emissions; 2020 (accessed on 24 April 2020)
7. Yang Z, Wang B, Jiao K. Life cycle assessment of fuel cell, electric and internal combustion engine vehicles under different fuel scenarios and driving mileages in China. *Energy* 2020, Article 117365
8. Messagie M, Boureima FS, Coosemans T, Macharis C, Van Mierlo JA. Range-Based Vehicle Life Cycle Assessment Incorporating Variability in the Environmental Assessment of Different Vehicle Technologies and Fuels. *Energies* 2014, 7(3), 1467–1482
9. Nordelöf A, Messagie M, Tillman AM, Söderman ML, Van Mierlo J. Environmental impacts of hybrid, plug-in hybrid, and battery electric vehicles—what can we learn from life cycle assessment?. *The International Journal of Life Cycle Assessment* 2014, 19, 1866–1890
10. The Guardian. https://www.theguardian.com/environment/2019/feb/12/electric-cars-already-cheaper-own-run-study; 2019 (accessed on 2 August 2021)
11. Colmenar-Santos A, Borge-Diez D, Ortega-Cabezas PM, Miguel-Camiña JV. Macro-economic impact, reduction of fee deficit and profitability of a sustainable transport model based on electric mobility. Case study: City of León (Spain). *Energy* 2014, 65, 303–318
12. Tarroja B, Li Z, Wifvat V, Shaffer B, Samuelsen S. Assessing the stationary energy storage equivalency of vehicle-to-grid charging battery electric vehicles. *Energy* 2016, 106, 673–690
13. Yang Z, Noori M, Tatari O. Vehicle to Grid regulation services of electric delivery trucks: Economic and environmental benefit analysis. *Applied Energy* 2016, 170, 161–175
14. Hofmann J, Guan D, Chalvatzis K, Hong H. Assessment of electrical vehicles as a successful driver for reducing CO_2 emissions in China. *Applied Energy* 2016, 184(15), 995–1003
15. Thiel C, Nijs W, Simoes S, Schmidt J, Van Zyl A, Schmid E. The impact of the EU car CO_2 regulation on the energy system and the role of electro-mobility to achieve transport decarbonisation. *Energy Policy* 2016, 96, 153–166
16. Colmenar-Santos A, de Palacio-Rodriguez C, Rosales-Asensio E, Borge-Diez D. Estimating the benefits of vehicle-to-home in islands: The case of the Canary Islands. *Energy* 2017, 134, 311–322
17. Noori M, Yang Z, Onat NC, Gardner S, Tatari O. Light-duty electric vehicles to improve the integrity of the electricity grid through Vehicle-to-Grid technology: Analysis of regional net revenue and missions savings. *Applied Energy* 2016, 168, 146–158
18. Darcovich K, Recoskie S, Ribberink H, Pincet F, Foissac A. Effect on battery life of vehicle-to-home electric power provision under Canadian residential electrical demand. *Applied Thermal Engineering* 2017, 114, 1515–1522
19. The International Council on Clean Transportation. https://theicct.org/sites/default/files/publications/EV-life-cycle-GHG_ICCT-Briefing_09022018_vF.pdf; 2018 (accessed on 15 July 2021)
20. Huang Y, Ng ECY, Zhou JL, Surawski NC, Chan EFC, Hong G. Eco-driving technology for sustainable road transport: A review. *Renewable and Sustainable Energy Reviews* 2018, 93, 596–609

21. Qi X, Wu G, Barth MJ, Boriboonsomsin K. Energy Impact of Connected Eco-driving on Electric Vehicles. In Road Vehicle Automation 4. 2nd Ed., Springer, New Yok, United States, 2017
22. HAL. https://hal.archives-ouvertes.fr/hal-01405291/document; 2016 (accessed on 12 June 2022)
23. Zhang C, Shen K, Yang F, Yuan C. Multiphysics Modeling of Energy Intensity and Energy Efficiency of Electric Vehicle Operation. *Procedia CIRP* 2019, 80, 322–327
24. Zeng W, Miwa T, Morikawa T. Eco-routing problem considering fuel consumption and probabilistic travel time budget. *Transportation Research Part D: Transport and Environment* 2020, 78, 102219
25. Wang J, Elbery A, Rakha HA. A real-time vehicle-specific eco-routing model for on-board navigation applications capturing transient vehicle behavior. *Transportation Research Part C: Emerging Technologies* 2019, 104, 1–21
26. Nunzio G, Thibault L, Sciarretta A.IEEE. International Conference on Intelligent Transportation Systems. https://ieeexplore.ieee.org/document/7795927?section=abstract; 2016 (accessed on 2 August 2021)
27. University of California. https://escholarship.org/uc/item/9z18z7xq; 2019 (accessed on 25 May 2022)
28. Kostopoulos ED, Spyropoulos GC, Kaldellis JK. Real-world study for the optimal charging of electric vehicles. *Energy Reports* 2020, 6, 418–426
29. Ko YD, Jang YJ. Efficient design of an operation profile for wireless charging electric tram systems. *Computers & Industrial Engineering* 2019, 127, 1193–1202
30. Here®. https://developer.here.com/; 2020 (accessed 29 March 2022)
31. Open Charge Map. https://openchargemap.org/site; 2020 (accessed 29 March 2021)
32. Alcala's town hall. http://traficoalcala.com/index.php?url=/estado/grafico; 2020 (accessed 29 April 2020)
33. Alcala's town hall. http://traficoalcala.com/index.php?url=/estado/tabla; 2019 (accessed 29 April 2021)
34. García-Pérez A. Estadística básica con R. 1st ed., UNED, Vizcaya, Spain, 2011
35. Government of Community of Madrid. http://gestiona.madrid.org/baco_web/html/web/AccionSeriesDestacadas.icm?acceso=8&annos=5; 2020 (accessed on 5 April 2020)
36. Alcala de Henares Town Hall. https://www.epdata.es/datos/datos-graficos-estadisticas-municipio/52/alcala-henares/582; 2020 (accessed on 5 April 2021)
37. Government of Community of Madrid. http://www.dgt.es/es/seguridad-vial/estadisticas-e-indicadores/parque-vehiculos/series-historicas/; 2019 (accessed on 5 April 2022)
38. International Energy Agency. https://www.iea.org/reports/global-ev-outlook-2019; 2019 (accessed on 4 March 2023)
39. EV volumes. https://www.ev-volumes.com/; 2019 (accessed on 4 March 2020)
40. International Council on Clean Transportation. https://theicct.org/publications/update-US-2030-electric-vehicle-cost; 2019 (accessed on 4 March 2020)
41. Spanish National Bank. https://www.bde.es/f/webbde/SES/AnalisisEconomico/AnalisisEconomico/ProyeccionesMacroeconomicas/ficheros/be08062020-proy.pdf; 2020 (accessed on 2 August 2020)

SPRINGER NATURE

GPSR Compliance

The European Union's (EU) General Product Safety Regulation (GPSR) is a set of rules that requires consumer products to be safe and our obligations to ensure this.

If you have any concerns about our products, you can contact us on ProductSafety@springernature.com

In case Publisher is established outside the EU, the EU authorized representative is:

Springer Nature Customer Service Center GmbH
Europaplatz 3
69115 Heidelberg, Germany

The manufacturer's authorised representative in the EU is Springer Nature Customer Service Centre GmbH, Europaplatz 3, 69115 Heidelberg, Germany. If you have any concerns regarding our products, please contact ProductSafety@springernature.com

Printed and bound by CPI Group (UK) Ltd, Croydon, CR0 4YY
28/09/2026
02224081-0002